Water-in-Plants Bibliography
volume 7 1981

References no. 8129–9676 / ABD–ZYA

Editors J. Pospíšilová and J. Solárová

Dr W. Junk bv Publishers The Hague/Boston/London 1982

Contributors

J. Solárová
J. Pospíšilová
Z. Šesták
J. Čatský
I. Tichá
D. Hodáňová
J. Zima

ISBN-13: 978-90-6193-907-8   e-ISBN-13: 978-94-009-8035-8
DOI: 10.1007/978-94-009-8035-8

PREFACE

The seventh volume of Water-in-Plants Bibliography includes papers in all fields of plant water relations research which appeared during the year 1981 – from theoretical considerations about the state of water in cells and its membrane transport to drought resistance of plants or physiological significance of irrigation. In addition to papers devoted entirely to plant water relations, papers on other topics are included if they contain data on plant hydration level, water vapour efflux, rate of water uptake or water transport, etc., or if they contain valuable methodological information (measurement of selected microclimatic factors, soil moisture etc.).

We have tried to cover fully the relevant papers which have been published in the most important scientific periodicals and books. Articles appeared in local journals, mimeographed booklets, abstracts of thesis and of symposia contributions, etc., were chosen mostly from reprints received directly from authors. The courtesy of those authors who have already supplied us with reprints and lists of their publications is highly appreciated. The manuscript is usually prepared in May and June of the year following the year which it covers. Unfortunately some reprints come later and thus the respective references appear in the following volume, with one year delay.

To maximize the value of the bibliography the references are arranged alphabetically according to the authors' names, and each volume is provided with three indexes. The authors' index contains all names of authors, co-authors and editors. Plant genera used as experimental material are indexed according to their Latin names. The subject index covers primary items chosen according to the interest of water relations researchers. Its preparation was based not only on the titles, key words and abstracts but also on the whole content of the article. By combining two or more items a more detailed information may be obtained.

Since more than 1500 relevant papers dealing with plant water relations and relative topics are published every year and included in this bibliography, and since all citations have been checked with the originals, collecting and preparing for publication such a large amount of material would have been impossible without the collaboration of our colleagues from the Department of Physiology of Photosynthesis and Water Relations of the Institute of Experimental Botany of the Czechoslovak Academy of Sciences in Prague. We have also acknowledge with thanks the cooperation of Mrs. Ludmila Hávová, Mrs. Lenka Kolčabová and Mrs. Irena Vaňková who helped in typing card material and the librarian of our Institute Mrs. Zora Zawoyská who helped us with checking the references.

Dr. Jana Pospíšilová and Ing. Jarmila Solárová

Institute of Experimental Botany
Czechoslovak Academy of Sciences

Flemingovo nám. 2
160 00 PRAHA 6
Czechoslovakia

Praha, 22 September 1982

INSTRUCTIONS FOR USE

All references are arranged alphabetically according to the authors' names. They are numbered and these numbers are used in the indexes. An asterisk preceding the number denotes the reference published in the preceding period (1975 - 1978).

Authors' names are presented in the spelling used in the original paper. If this spelling does not correspond to the spelling usually used by the author (e.g. Russian papers of English authors), one spelling is referred to the other in the Authors' Index. Like the transcriptions they are alphabetically arranged mostly according to the authors' own references. Nevertheless, the editors apologize for some misinterpretations which are partly corrected by the cross-indexing in the Authors' Index.

The references contain the original unshortened title of the paper (book). English, French, and German titles are cited in the original language. Titles in other languages are supplemented with a translation in English (using the title of the respective English abstract, if it is presented). Titles of Japanese, Chinese etc. papers are given in English translation only. In both these cases the abbreviations of the original language and the language of the abstract are given in brackets at the end of the reference. The following abbreviations are used most frequently:

Belorussian	Japanese
Bulgarian	Latvian
Chinese	Lithuanian
Croatian	Norwegian
Danish	Polish
English	Russian
Esthonian	Roumanian
French	Slovak
German	Spanish
Georgian	Swedish
Hungarian	Ukrainian
Italian	Uzbeg

The transliteration of Cyrillic characters is in accordance with the BSI-ASA/SC-Z39 draft table, i.e.:

a	а		p	п
b	б		r	р
ch	ч		s	с
d	д		sh	ш
e	е		shch	щ
ė	э		t	т
f	ф		ts	ц
g	г		u	у
i	и		v	в
ï	й		y	ы
k	к		ya	я
kh	х		yu	ю
l	л		z	з
m	м		zh	ж
n	н		"	ъ
o	о		'	ь

Several exceptions apply for Ukrainian and Belorussian:

Ukrainian:	y	и
	i	i
	ï	ï
Belorussian:	ŭ	ў

The journals' names are abbreviated mainly according to the Style Manual for Biological Journals (Second Edition, Amer. Institute of Biological Sciences, Washington, D.C. 1964), e.g.:

Abhandlungen
Abstract
Abteilung
Academy
Acker
Acta
Advances
Africa (-ican)
agricultural
Agriculture
Agrobiology (-ogiya)
Agrobotanica
Agrokēmia
Agronomy
agropecuaria
Akademie (-emiya)
Algology
allgemeine
Amēlioration
America
American
Anais (-alele)
Analysis
analytical
Anatomy
angewandte
animal
Annales (-als)
annual
anorganic (-anisch)
applied
aquatic
Arbeit
Archiv
Argentina
Association
Atmosphere
atmospheric
atomic
Australia (-ralian)
Azerbaĭdzhanskaya
Bacteriology
Beiheft
Beiträge
Belgique
Belorusskaya
Berichte
biochemical
Biochemie
Biochemistry
biochimica
biokhimicheskiĭ
Biokhimiya
Bioklimatologie
Biologia (-ogy)
biological (-ogisk)
biophysical
Biophysics
Bodenkunde
Boletin (-ettino)
Bolgarskiĭ
botanica (-anicorum)
botanical (-anisca)
Botanika (-any)

Brasileira
Brazil
Breeding
British
Bulletin (-etins)
Byulleten
California
Canada (-adian)
cellular (-ulaire)
Center
central
Centralblatt
Československý
chemical
Chemistry
chimicus
Chinese
Chromatography
Chronicle
Ciencia
cientificas
College
Commision
Communication
comparative
Comptes Rendus
Conference
Congress
Conservation
Contamination
Contribution
Control
Croatica
cultural
Culture
current
Cytobiology
Cytochemistry
Cytology
Czechoslovak
Danske
dendrological
Dendrology
Department
Deutsche (-schland)
Development
Disease
Dissertation
Division
Doklady
Dopovidi
Drainage
ecological
Ecology
Economy
Edafology
Education
Ėkologiya
ėksperimental'nyĭ
Embryology
Encyclopedia
Engineering
Enology
Entomology

environmental
Enzymology
Estonskaya
European
Experiment
experimental
Faculty
Federation
Fizika
Fiziologiya
Flurbereinigung
forestiere
Forestry
Forschung
Foundation
France
Gazette
general
genetical
geneticheskiĭ
Genetics (-ika)
Geobotany
Geofizika
Geophysics
Gesellschaft
Giornale
gosudarstvennyĭ
Government
Grassland
Gruzinskaya
Helveticus
Histochemistry
Histoire (-ory)
Histology
horticultural
Horticulture
Hungaricae
Hungaricus
Husbandry
Hydrobiology
Hydrology
Indian
Industry
inorganic
Institute
Institutului
international
Investigation
Irrigation
Isotopes
issledovatel'skiĭ
Italian (-y)
Izvestiya
Jahrbuch
Japan (-anese)
Journal
Khimiya
Klasse
Kongelige
Közlemenyek
kul'turnykh
Laboratory
Landbauforschung
Landwirtschaft

lesní (-ího)
Letters
Limnology
Linnean
Litovskoĭ
Lucrarile
Magazin
Management
marina (-ine)
Material
Mathematics
Mededelingen
mediterranean
Meldinger
Meteorology
Microbiology
Midland
Mitteilungen
Modeling
modern
molecular
Monographiae (-aphy)
Moskovskiĭ (-ovskogo)
Mycology
national
natural
Naturalist
naturelle
naturkundliche
Naturforschung
nauchnye (-nyĭ)
Neerlandica
Netherland
New Zealand
Norges
Norwegian
Notiser
nuclear
Nutrition
obshcheĭ (-iĭ)
Oceanography
Oecologia
Ökologie
Optics
opytnaya (-yĭ)
organic
original
ornamental
Otdelenie
Paleobotany
Palynology
Pathology
pedagogicheskiĭ
Pesquisa
Pesticide
Pflanzen-
Pflanzenernährung
Pflanzenphysiologie
Pflanzenzüchtung
Philosophy
Photogrammetric
Phycology
physical
Physics

physiological
Physiology
Phytologist
Phytopathology
Phytotaxonomy
Plantarum
Polonica (-ska)
Pollution
Práce
Practice
prikladnoĭ
Proceedings
Progress
Publication
Publishers
Quality
quantitative
Quarterly
Radiation
Radiobiology
Rasteniĭ
Rastenievodstvo
Recherche (-erches)
Report
Research
Resources
Review (-ista, -ue)
Rivista
Roczniky

rolniczych
Rostlin (-lina)
rostlinná
Roumaine
royal
Russian
Russkiĭ
Sbornik
Scandinavica
Scandinavicus
School
Science
scientific
Section
Selektsiya
Selskabs
Sel'skokhozyaĭstvo
Series (-iya)
Service
Shkoly (-oly)
Sibirskiĭ (-skogo)
Skrifter
Slovak (-enská)
Society
Soobshcheniya
Sovetskiĭ (-iet)
sovremennyĭ
special
sperimentale

SSSR
Stantsii (-ntsiya)
Station
stiintifice
subtropicale
summary
Supplement
Survey
Swedish
Symposium
System
Tagungsberichte
technical (-nische)
Technology
Tekhnika
theoretical
thermal
Tidsskrift
Tijdschrift
Toxicology
Transactions
Travail (-aux)
tropical (-icale)
Trudy
Turkmenskaya
uchenye
Ugeskrift
United Kingdom
Ukraĭnian

Ukrains'kaya
Universidad (-ersity)
US, USA
USSR
Uzbekskiĭ (-ekskaya)
vědecké (-ecký)
vegetable
végétale
Verhandlungen
Veröffentlichungen
Vestnik
Videnskabernes
Virology
Virusforschungen
Viticulture
Volume
Voprosy
vostochnyĭ
vsesoyuznyĭ
vyssheĭ (-iĭ)
výzkumný (-umného)
Weekblatt
Wetenschappen
Wissenschaft
Zapiski
Zeitschrift
Zeitung
Zentralblatt
Zhurnal

The numbers at the end of each reference of a journal article denote: volume (issue) : first page - last page, year of publication. The number of issue is given only in journals where each issue is paginated separately.

Book titles are cited according to the title page, not to the book jacket or cover. The publishing house, place and year of publication are included.

Printers' errors in the original papers are marked by underlining the respective words (letters).

*8129 - ABDALLAH, M.M., KHALIFA, M.A.: Effect of drought conditions on some chemical constituents of cotton seedlings. - Z. Acker- Pflanzenbau 149: 424-429, 1980.

8130 - ABDULRAHMAN, F.S., WILLIAMS, G.J.,III: Temperature and salinity regulation of growth and gas exchange of *Salicornia fructicosa* (L.) L. - Oecologia 48: 346 -352, 1981.

*8131 - ABOU-HAIDAR, S.S., FERERES, E., HARRIS, R.W.: Drought adaptation of two species of *Cotoneaster*. - J. hort. Sci. 55: 219-227, 1980.

*8132 - ACKERSON, R.C.: Osmoregulation and water stress adaptation in cotton. - Plant Physiol. 65 (Suppl.): 154, 1980.

8133 - ACKERSON, R.C.: Osmoregulation in cotton in response to water stress.II. Leaf carbohydrate status in relation to osmotic adjustment. - Plant Physiol. 67: 489-493, 1981.

8134 - ACKERSON, R.C., HEBERT, R.R.: Osmoregulation in cotton in response to water stress.I. Alterations in photosynthesis, leaf conductance, translocation, and ultrastructure. - Plant Physiol. 67: 484-488, 1981.

8135 - ACOCK, B., GRANGE, R.I.: Equilibrium models of leaf water relations. - In: ROSE, D.A., CHARLES-EDWARDS, D.A. (ed.): Mathematics and Plant Physiology. Pp. 29-47. Academic Press, London - New York - Toronto - Sydney - San Francisco 1981.

8136 - ADEOYE, K.B., RAWLINS, S.L.: A split-root technique for measuring root water potential. - Plant Physiol. 68: 44-47, 1981.

*8137 - ADJEI, G.B., KIRKHAM, M.B.: Evaluation of winter wheat cultivars for drought resistance. - Euphytica 29: 155-160, 1980.

8138 - AFANAS'EV, V.P., MALOFEEV, V.M.: Vozrastnaya reaktsiya khlopchatnika na ste- pen' mineralizatsii orositel'noĭ vody. [Ageing of cotton plants affected by mineralization of irrigation water.] - Sel'skokhoz. Biol. 16: 77-79, 1981. [In R, ab: E.]

8139 - AGARWAL, M.C., GOEL, A.C.: Effect of field levelling quality on irrigation efficiency and crop yield - Agr. Water Manage. 4: 457-464, 1981.

8140 - AGGARWAL, P.K., CHATURVEDI, G.S., SINHA, S.K.: Water potential in relation to stomatal resistance and leaf growth in some cereals. - Ind. J. exp. Biol. 19: 882-884, 1981.

8141 - AGGARWAL, P.K., CHATURVEDI, G.S., SINHA, S.K.: Day time changes in chlorophyll content in plants at high temperature & radiation. - Ind. J. exp. Biol. 19: 1127-1130, 1981.

*8142 - ÅGREN, G.I.: Problems involved in modelling tree growth. - Studia forest. Suecica 160 (LINDER, S. (ed.): Understanding and Predicting Tree Growth.): 7-18, 1981.

*8143 - ÅGREN, G.I., AXELSSON, B.: PT - a tree growth model. - Ecol. Bull. (Stock- holm) 32 (PERSSON, T. (ed.): Structure and Function of Northern Coniferous Forests - An Ecosystem Study.): 525-536, 1980.

*8144 - AHMAD, R., ABDULLAH, Z.: Salinity induced changes in the growth and chemical composition of potato. - Pak. J. Bot. 11: 103-112, 1979.

*8145 - AHMAD, R., ABDULLAH, Z.: Saline agriculture under desert conditions. - In: BISHAY, A., McGINNIES, W.G. (ed.): Advances in Desert and Arid Land Technolo- gy and Development. Volume 1. Pp. 593-618. Harwood Academic Publishers, Chur - London - New York 1979.

*8146 - AHMED, B., QUILT, P.: Effect of soil moisture stress on yield, nodulation and nitrogenase activity of *Macroptilium atropurpureum* cv. Siratro and *Desmodium intortum* cv. Greenleaf. - Plant Soil 57: 187-194, 1980.

*8147 - AKESON, W.R., HENSON, M.A., FREYTAG, A.H. WESTFALL, D.G.: Sugarbeet fruit germination and emergence under moisture and temperature stress. - Crop Sci. 20: 735-739, 1980.

8148 - AKSYONOV, S.I.: Regulatory role of water in biological systems. - Studia biophys. 85: 5-6, 1981.

8149 - AKSYONOV, S.I., GORYACHEV, S.N., NIKOLAEV, G.M.: Water states in low-humidity organisms. - Studia biophys. 85: 15-16, 1981.

8150 - AKUSHIE, P-L., CLERK, G.C.: Effect of relative humidity on viability of *Rhizopus oryzae* sporangiospores. - Trans. Brit. mycol. Soc. 76: 332-334, 1981.

*8151 - ALBERT, R., FALTER, J.: Stoffwechselphysiologische Untersuchungen an Blättern streusalzgeschädigter Linden in Wien. I. Salzgehalt und Ionenbilanz. - Phyton 18: 173-197, 1978.

*8152 - ALBERT, R., FALTER, J.: Stoffwechselphysiologische Untersuchungen an Blättern streusalzgeschädigter Linden in Wien. II. Stickstoff- und Kohlenhydratstoff-wechsel. - Phyton 19: 141-162, 1979.

8153 - ALBRECHT, S.L., BENNETT, J.M., QUESENBERRY, K.H.: Growth and nitrogen fixation of *Aeschynomene* under water stressed conditions. - Plant Soil 60: 309-315, 1981.

8154 - ALESSI, J., POWER, J.F., ZIMMERMAN, D.C.: Effects of seeding date and population on water-use efficiency and safflower yield. - Agron. J. 73: 783-787, 1981.

*8155 - ALEXANDER, J.M., LUCAS, W.J.: The effect of turgor pressure on ion transport in *Chara carollina.* - Plant Physiol. 65 (Suppl.): 148, 1980.

*8156 - ALHADEFF, M., LÜTZ, C., KLEIN, S., SCHIFF, J.A.: Influence of hydration on spectroscopic forms of protochlorophyll(ide) and chlorophyll(ide) in freeze--dried fractions from oat proplastids. - Plant Physiol. 65 (Suppl.): 66, 1980.

*8157 - ALINA, B.A., KLYSHEV, L.K.: Deĭstvie zasoleniya sredy na tsiklicheskoe fotofosforilirovanie v khloroplastakh risa. [Effect of medium salinization on cyclic photophosphorylation in rice chloroplasts.] - Izv. Akad. Nauk Kaz. SSR, Ser. biol. 1979(4): 10-14, 1979. [In R, ab: Kaz.]

8158 - ALLAWAY, W.G.: Anions in stomatal operation. - In: JARVIS, P.G., MANSFIELD, T.A. (ed.): Stomatal Physiology. Pp. 71-85. Cambridge University Press, Cambridge - London - New York - New Rochelle - Melbourne - Sydney 1981.

8159 - ALLAWAY, W.G., PITMAN, M.G., STOREY, R., TYERMAN, S.: Relationships between sap flow and water potential in woody or perennial plants on islands of the Great Barrier Reef. - Plant Cell Environ. 4: 329-337, 1981.

8160 - ALLEN, M.F., SMITH, W.K., MOORE, T.S.,Jr., CHRISTENSEN, M.: Comparative water relations and photosynthesis of mycorrhizal and non-mycorrhizal *Bouteloua gracilis* H.B.K. Lag ex Steud. - New Phytol. 88: 683-693, 1981.

8161 - ALLEN, M.J.: The temperature-dependent behaviour of a semiconductive property of plant membranes. - J. exp. Bot. 32: 855-859, 1981.

8162 - ALLISON, E.M., WALSBY, A.E.: The role of potassium in the control of turgor pressure in a gas-vacuolate blue-green alga. - J. exp. Bot. 32: 241-249, 1981.

8163 - ALONI, B., PRESSMAN, E.: Stem pithiness in tomato plants: The effect of water stress and the role of abscisic acid. - Physiol. Plant. 51: 39-44, 1981.

8164 - ALONI, B., PRESSMAN, E., SHAKED, R.: Effects of salts on NADH-malate dehydrogenase activity in celery leaves during their maturation and senescence. - Plant Sci. Lett. 20: 239-250, 1981.

*8165 - AL-SAIDI, I.: Studies on the influence of different concentrations of sodium chloride and calcium chloride salts on the growth of some grapevine cultivar transplants. - Mesopotamia J. Agr. 15: 125-135, 1980.

8166 - AMBRUS, L., ANTAL, E., KARSAI, H.A.: New electronic evaporation and rain measuring equipment. - Agr. Meteorol. 25: 35-43, 1981.

8167 - AMUNDSON, R.G., WEINSTEIN, L.H.: Joint action of sulfur dioxide and nitrogen dioxide on foliar injury and stomatal behavior in soybean. - J. environ. Qual. 10: 204-206, 1981.

8168 - ANDEREGG, B.N., LICHTENSTEIN, E.P.: A comparative study of water transpiration and the uptake and metabolism of [^{14}C]phorate by C_3 and C_4 plants. - J. agr. Food Chem. 29: 733-738, 1981.

8169 - ANDERSEN, K., ANDERSEN, S.: Increase in dry matter and decrease in moisture content during ripening of barley. - Acta agr. Scand. 31: 70-74, 1981.

*8170 - ANDERSSON, F.: Ecosystem research within the Swedish Coniferous Forest Project. - Ecol. Bull. (Stockholm) 32 (PERSSON, T. (ed.): Structure and Function of Northern Coniferous Forests - An Ecosystem Study.): 11-23, 1980.

*8171 - ANDERSSON, L.-Å.: Water transport in hardened and non-hardened seedlings of Scots pine. - Ecol. Bull. (Stockholm) 32 (PERSSON, T. (ed.): Structure and Function of Northern Coniferous Forests - An Ecosystem Study.): 215-218, 1980.

*8172 - ANDONOV, D.: Vzaimodeĭstvie mezhdu nachinite na napoyavane, polivniya rezhim i ravnishcheto na torene i vliyanieto im v"rkhu dobiva na zakharno tsveklo. [Interaction between type of irrigation, irrigation regime and fertilization and their effect on sugar beet yield.] - Rasteniev. Nauki 17(2): 82-91, 1980. [In Bulg, ab: R, F.]

*8173 - ANDONOV, D., V"LEV, V.: Poliven rezhim na zakharno tsveklo, otglezhdano na izluzhen chernozem-smolnitsa. [Irrigation regime of sugar beet grown on leached chernozem-smolnitsa soil.] - Rasteniev. Nauki 17(6): 80-91, 1980. [In Bulg, ab: R, F.]

8174 - ANDREWS, C.J., POMEROY, M.K.: The effect of flooding pretreatment on cold hardiness and survival of winter cereals in ice encasement. - Can. J. Plant Sci. 61: 507-513, 1981.

*8175 - ANDREWS, K.L., LA PRÉ, L.F.: Effects of Pacific spider mite on physiological processes of almond foliage. - J. econ. Entomol. 72: 651-654, 1979.

8176 - AONO, H., YANASE, Y., TANAKA, S.: [Studies on the development and distribution of roots and its soil conservation faculty in tea plants. IV. Influences of subsoil improvement and irrigation on the development of tea roots.] - Jap. J. Crop Sci. 50: 276-281, 1981. [In Jap, ab: E.]

8177 - APEL, P., SCHALLDACH, I.: Sortenunterschiede in der Photosynthese bei Gerste (Hordeum vulgare L.): CO$_2$-Abhängigkeit, Aktivität und Menge der Ribulose-1,5-bisphosphat-Carboxylase. - Biochem. Physiol. Pflanz. 176: 454-462, 1981.

8178 - **APELBAUM, A., YANG, S.F.:** Biosynthesis of stress ethylene induced by water deficit. - Plant Physiol. 68: 594-596, 1981.

*8179 - **ARONSSON, A., ELOWSON, S.:** Effects of irrigation and fertilization on mineral nutrients in Scots pine needles. - Ecol. Bull. (Stockholm) 32 (PERSSON, T. (ed.): Structure and Function of Northern Coniferous Forests - An Ecosystem Study.): 219-228, 1980.

*8180 - **ARORA, B.B., LAMBA, L.C.:** Stomata in the pericarp of *Brassica oleracea* var. *botrytis* Linn. and *Eruca sativa* Mill. - Proc. Ind. Acad. Sci., Sect. B (Plant Sci.) 89: 23-28, 1980.

*8181 - **ASLYNG, H.C.:** Solenergi, vand, jord og planteproduktion. [Sun energy, wind, soil and plant production.] - BP Nyhedstjeneste 30(91): 6-9, 1979. [In: Dan.]

8182 - **ASPINALL, D., PALEG, L.G.:** Proline accumulation: Physiological aspects. - In: PALEG, L.G., ASPINALL, D. (ed.): The Physiology and Biochemistry of Drought Resistance in Plants. Pp. 205-241. Academic Press, Sydney - New York - London - Toronto - San Francisco 1981.

*8183 - **ATANASIU, N., THIAGALINGAM, K.:** A preliminary germination and growth test of some Malaysian and exotic paddy varieties on salinity tolerance. - Malaysian agr. J. 51: 250-254, 1978.

*8184 - **AUCLAIR, D., MÉTAYER, S.:** Méthodologie de l'évaluation de la biomasse aérienne sur pied et de la production en biomasse des taillis. - Acta Oecol.-Oecol. appl. 1: 357-377, 1980.

8185 - **AUST, H.-J.:** Über den Verlauf von Mehltauepidemien innerhalb des Agro-Ökosystems Gerstenfeld (Acta Phytomedica H. 7). - Verlag Paul Parey, Berlin - Hamburg 1981.

8186 - **AUSTIN, R.B., MORGAN, C.L., FORD, M.A.:** A field study of the carbon economy of normal and 'topless' field beans (*Vicia faba*). - In: THOMPSON, R. (ed.): *Vicia faba*. Physiology and Breeding. Pp. 60-77. Martinus Nijhoff, The Hague 1981.

*8187 - **AVITA, S., INAMDAR, J.A.:** Structure and ontogeny of stomata in *Ranunculaceae* and *Paeoniaceae*. - Flora 170: 354-370, 1980.

8188 - **AYRES, P.G.:** Root growth and solute accumulation in pea in response to water stress and powdery mildew. - Physiol. Plant Pathol. 19: 169-180, 1981.

8189 - **AYRES, P.G.:** Powdery mildew stimulates photosynthesis in uninfected leaves of pea plants. - Phytopathol. Z. 100: 312-318, 1981.

8190 - **AYRES, P.G.:** Responses of stomata to pathogenic microorganisms. - In: JARVIS, P.G., MANSFIELD, T.A. (ed.): Stomatal Physiology. Pp. 205-221. Cambridge University Press, Cambridge - London - New York - New Rochelle - Melbourne - Sydney 1981.

8191 - **BAAS, P.:** A note on stomatal types and crystals in the leaves of Melastomataceae. - Blumea 27: 475-479, 1981.

8192 - **BABALOLA, O., OPUTA, C.:** Effects of planting patterns and population on water relations of maize. - Exp. Agr. 17: 97-104, 1981.

8193 - **BAČA, F., VASIĆ, G.:** Prilog proučavanju uticaja navodnjavanja, đubrenja i gustine biljaka na prinos nekih hibrida kukuruza i intenzitet napada kukuruznog plamenca (*Ostrinia nubilalis* Hbn.). [A contribution to the study of the influence of irrigation, fertilization and plant density on yield and intensity of infestation by the european corn borer (*Ostrinia nubilalis* Hbn.) in some maize hybrids.] - Arhiv poljopr. Nauke 42: 93-101, 1981. [In Croat, ab: E.]

8194 - **BADANOVA, K.A., BALINA, N.V.**: Povyshenie produktivnosti podsolnechnika metodom predposevnogo zakalivaniya k zasukhe. [Increase of sunflower yield by the
method of presowing hardening to drought.] - Fiziol. Rast. 28: 1069-1071,
1981. [In R.]

8195 - **BAIER, W.**: Water balance in crop-yield models. - In: BERG, A. (ed.): Application of Remote Sensing to Agricultural Production Forecasting. Pp. 119-129.
A.A. Balkena, Rotterdam 1981.

8196 - **BAILEY, H.P.**: Climatic features of deserts. - In: EVANS, D.D., THAMES, J.L.
(ed.): Water in Desert Ecosystems. Pp. 13-41. Dowden, Hutchinson & Ross, Inc.,
Stroudsburg 1981.

8197 - **BAILEY, W.G., DAVIES, J.A.**: Bulk stomatal resistance control on evaporation. -
Boundary-Layer Meteorol. 20: 401-415, 1981.

8198 - **BAILEY, W.G., DAVIES, J.A.**: Evaporation from soybeans. - Boundary-Layer
Meteorol. 20: 417-428, 1981.

8199 - **BAKER, E.A., HUNT, G.M.**: Developmental changes in leaf epicuticular waxes in
relation to foliar penetration. - New Phytol. 88: 731-747, 1981.

8200 - **BAKRADZE, N.G., BALLA, Yu.I., METREVELI, I.M.**: O vozmozhnom mekhanizme pro-
tsessa kristallizatsii vody v tkanyakh rasteniĭ. [Possible mechanism of water
cristallization in plant tissues.] - Biofizika 26: 719-723, 1981. [In R,
ab: E.]

8201 - **BAKRADZE, N.G., MOISTSRAPISHVILI, K.M., KESHELASHVILI, L.V.**: O protsesse
kristallizatsii vody v tkanyakh rasteniĭ. [Water crystallization in plant
tissues.] - Biofizika 26: 119-123, 1981. [In R, ab: E.]

8202 - **BALASUBRAMANIAN, V., YAYOCK, J.Y.**: Effect of gypsum and moisture stress on
growth and pod-fill of groundnut (*Arachis hypogaea* L.). - Plant Soil 62:
209-219, 1981.

8203 - **BALDOCCHI, D.D., VERMA, S.B., ROSENBERG, N.J.**: Mass and energy exchanges of
a soybean canopy under various environmental regimes. - Agron. J. 73: 706-
710, 1981.

8204 - **BALDOCCHI, D.D., VERMA, S.B., ROSENBERG, N.J.**: Seasonal and diurnal variation in the CO_2 flux and CO_2-water flux ratio of alfalfa. - Agr. Meteorol.
23: 231-244, 1981.

8205 - **BALDOCCHI, D.D., VERMA, S.B., ROSENBERG, N.J.**: Environmental effects on the
CO_2 flux and CO_2-water flux ratio of alfalfa. - Agr. Meteorol. 24: 175-184,
1981.

8206 - **BANBA, H., OHKUBO, T.**: [Relationship between root distribution of upland
crops and their yield. III. Influence of soil moisture levels on root
distribution and root dry matter of upland-cultured paddy rice, crossbred
rice of paddy rice and upland rice, and upland rice.] - Jap. J. Crop Sci.
50: 1-7, 1981. [In Jap, ab: E.]

8207 - **BANBA, H., OHKUBO, T.**: Relationship between root distribution of upland
crops and their yield. IV. Soil water uptake pattern in upland rice under
different soil moisture levels. - Jap. J. Crop Sci. 50: 176-180, 1981.

8208 - **BARANOVA, M.A.**: O laterotsitnom tipe ust'ichnogo apparata u tsvetkovykh.
[On the laterocytic type of the stomatal apparatus in *Angiospermae*.] - Bot.
Zh. 66: 179-186, 1981. [In R, ab: E.]

*8209 - **BARAVIKOVA, A.M.**: Ab vodnym rėzhyme sasny zvychaĭnaĭ na ŭmovakh pramyslovaga
npadpryemstva. [Water regime of common pine under conditions of industrial
enterprises.] - Vestsi Akad. Navuk Belarus. SSR, Ser. biyal. Navuk 1980 (6):
16-19, 138, 1980. [In Belorus, ab: R, E.]

8210 - **BARBER, J., MALKIN, S.**: Salt-induced microscopic changes in chlorophyll
fluorescence distribution in the thylakoid membrane. - Biochim. biophys.
Acta 634: 344-349, 1981.

*8211 - **BARKER, A.V.**: Nutritional factors in photosynthesis of higher plants. -
J. Plant Nutr. 1: 309-342, 1979.

*8212 - **BARRAN, L.R.**: Effect of heat, freeze-thawing and desiccation on the survival
of *Fusarium sulphureum* spores. - Trans. Brit. mycol. Soc. 75: 425-427, 1980.

*8213 - **BARRICK, W.E., DAVIDSON, H.**: Deicing salt spray injury in Norway maple as
influenced by temperature and humidity treatments. - HortScience 15: 203-205,
1980.

*8214 - **BARTA, A.L.**: Regrowth and alcohol dehydrogenase activity in waterlogged
alfalfa and birdsfoot trefoil. - Agron. J. 72: 1017-1020, 1980.

8215 - **BARTHLOTT, W., SCHILL, R.**: Oberflächenskulpturen bei höheren Pflanzen. -
Prog. Bot. 43: 27-38, 1981.

*8216 - **BARUCH, Z.**: Elevational differentiation in *Espeletia schultzii* (Compositae),
a giant rosette plant of the Venezuelan paramos. - Ecology 60: 85-98, 1979.

8217 - **BATAL, K.M., SMITTLE, D.A.**: Response of bell pepper to irrigation, nitrogen,
and plant population. - J. Amer. Soc. hort. Sci. 106: 259-262, 1981.

*8218 - **BATANOUNY, K.H.**: Water economy of desert plants. - In: HALAZI-KUN, G.J. (ed.):
Pollution and Water Resources. Volume XIII, Part 2. Pp. 167-177. Pergamon
Press, New York - Oxford - Toronto - Sydney - Paris - Frankfurt 1980.

8219 - **BATANOUNY, K.H.**: Eco-physiological studies on desert plants. X. Contribution
to the autecology of the desert chasmophyte *Stachys aegyptiaca* Pers. -
Oecologia 50: 422-426, 1981.

8220 - **BATANOUNY, K.H., EBEID, M.M.**: Diurnal changes in proline content of desert
plants. - Oecologia 51: 250-252, 1981.

8221 - **BATES, L.M., HALL, A.E.**: Stomatal closure with soil water depletion not
associated with changes in bulk leaf water status. - Oecologia 50: 62-65,
1981.

8222 - **BAUDER, J.W., ENNEN, M.J.**: Soil water use efficiency of sunflowers. - North
Dakota Farm Res. 39: 9-12, 1981.

8223 - **BAUMEISTER, W., MERTEN, A.**: Einfluss der NaCl-Konzentration in der Nährlösung
auf das Wachstum und die Wurzelanatomie bei zwei Unterarten von *Festuca rubra*
L. - Angew. Bot. 55: 401-408, 1981.

8224 - **BEADLE, C.L., NEILSON, R.E., JARVIS, P.G., TALBOT, H.**: Photosynthesis as
related to xylem water potential and carbon dioxide concentration in Sitka
spruce. - Physiol. Plant. 52: 391-400, 1981.

*8225 - **BEALL, P.T.**: Water-macromolecular interactions during the cell cycle. - In:
WHITSON, G.L. (ed.): Nuclear-Cytoplasmic Interactions in the Cell Cycle. Pp.
223-247. Academic Press, New York - San Francisco - London 1980.

8226 - BECWAR, M.R., RAJASHEKAR, C., HANSEN BRISTOW, K.J., BURKE, M.J.: Deep under-cooling of tissue water and winter hardiness limitations in timberline flora. - Plant Physiol. 68: 111-114, 1981.

*8227 - BEDENKO, V.P.: Fotosintez i Produktivnost' Pshenitsy na Yugo-Vostoke Kazakh-stana. [Photosynthesis and Productivity of Wheat in the South-East of Kazakhstan.] - Nauka Kaz. SSR, Alma-Ata 1980.[In R.]

8228 - BEDUNAH, D., TRLICA, M.J.: Carbon dioxide exchange of ponderosa pine as affected by sodium chloride and polyethylene glycol. - Forest Sci. 27: 139-146, 1981.

8229 - BELFORD, R.K.: Response of winter wheat to prolonged waterlogging under out-door conditions. - J. agr. Sci. 97: 557-568, 1981.

*8230 - BELIĆ, J. (ed.): Fiziologija Kukuruza. [Physiology of Maize.] - Srpska Akad. Nauka Umet., Beograd 1980. [In Serb, ab: E.]

8231 - BELL, C.J., INCOLL, L.D.: A handpiece for the simultaneous measurement of photosynthetic rate and leaf diffusive conductance. I. Design. - J. exp. Bot. 32: 1125-1134, 1981.

8232 - BELL, C.J., INCOLL, L.D.: A handpiece for the simultaneous measurement of photosynthetic rate and leaf diffusive conductance. II. Calibration. - J. exp. Bot. 32: 1135-1142, 1981.

8233 - BELL, C.J., SQUIRE, G.R.: Comparative measurements with two water vapour diffusion porometers (dynamic and steady-state). - J. exp. Bot. 32: 1143-1156, 1981.

*8234 - BELL, K.R., BLANCHARD, B.J., SCHMUGGE, T.J., WITCZAK, M.W.: Analysis of sur-face moisture variations within large-field sites. - Water Resour. Res. 16: 796-810, 1980.

*8235 - BELOKOBYL'SKIĬ, I.M.: Vliyanie makro- i mikroélementov na zasukhoustoĭchivost' gibridov kukuruzy v usloviyakh Dombassa. [Effect of macro- and microelements on drought resistance of maize hybrids in Dombass.] - In: Mineral'noe Pitanie i Produktivnost' Rasteniĭ. Pp. 242-247. Naukova Dumka, Kiev 1978. [In R.]

*8236 - BENECKE, P., VAN DER PLOEG, R.R.: Das hydrologische Verhalten ungesättigter Bodenschichten am Beispiel forstlicher Ökosysteme. - Z. Pflanzenernähr. Bodenk. 142: 169-184, 1979.

8237 - BENECKE, P., VAN DER PLOEG, R.R.: The soil environment. - In: LANGE, O.L., NOBEL, P.S., OSMOND, C.B., ZIEGLER, H. (ed.): Physiological Plant Ecology I. Responses to the Physical Environment. Pp. 539-559. Springer-Verlag, Berlin - Heidelberg - New York 1981.

8238 - BENECKE, U., SCHULZE, E.-D., MATYSSEK, R., HAVRANEK, W.M.: Environmental control of CO_2-assimilation and leaf conductance in *Larix decidua* Mill. I. A comparison of contrasting natural environments. - Oecologia 50: 54-61, 1981.

*8239 - BENGTSON, C.: Effects of water stress on Scots pine. - Ecol. Bull. (Stockholm) 32 (PERSSON, T. (ed.): Structure and Function of Northern Coniferous Forests - An Ecosystem Study.): 205-213, 1980.

8240 - BENNETT, J.H.: Photosynthesis and gas diffusion in leaves of selected crop plants exposed to ultraviolet-B radiation. - J. environ. Qual. 10: 271-275, 1981.

8241 - BENNETT, J.M., BOOTE, K.J., HAMMOND, L.C.: Alterations in the components of peanut leaf water potential during desiccation. - J. exp. Bot. 32: 1035-1043, 1981.

8242 - BENNETT, J.M., SULLIVAN, C.Y.: Effects of water stress preconditioning on net photosynthetic rate of grain sorghum. - Photosynthetica 15: 330-337, 1981.

8243 - BERNIER, G., KINET, J.-M., SACHS, R.M.: The Physiology of Flowering. Volume I. The Initiation of Flowers. - CRC Press, Inc., Boca Raton 1981.

*8244 - BERSHTEĬN, B.I., OKANENKO, A.S.: Regulyatornaya rol' kaliya v fotosinteze i produktivnosti rasteniĭ. [Regulatory role of potassium in photosynthesis and plant productivity.] - In: Mineral'noe Pitanie i Produktivnost' Rasteniĭ. Pp. 159-167, 323. Naukova Dumka, Kiev 1978. [In R.]

8245 - BEWLEY, J.D.: Protein synthesis. - In: PALEG, L.G., ASPINALL, D. (ed.): The Physiology and Biochemistry of Drought Resistance in Plants. Pp. 261-282. Academic Press, Sydney - New York - London - Toronto - San Francisco 1981.

8246 - BEWLEY, J.D. (ed.): Nitrogen and Carbon Metabolism. - Martinus Nijhoff / Dr. W. Junk Publishers, The Hague - Boston - London 1981.

8247 - BHARDWAJ, R., SINGHAL, G.S.: Effect of water stress on photochemical activity of chloroplasts during greening of etiolated barley seedlings. - Plant Cell Physiol. 22: 155-162, 1981.

8248 - BIDDINGTON, N.L.: Thermodormancy and the prevention of desiccation injury in celery seeds. - Ann. appl. Biol. 98: 558-562, 1981.

*8249 - BIGGS, A.R., DAVIS, D.D.: Stomatal response of three birch species exposed to varying acute doses of SO_2. - J. Amer. Soc. hort. Sci. 105: 514-516, 1980.

8250 - BIGGS, A.R., DAVIS, D.D.: Sulfur dioxide injury, sulfur content, and stomatal conductance of birch foliage. - Can. J. Forest Res. 11: 69-72, 1981.

8251 - BIKHELE, Z., MOLDAU, Kh., ROSS, Yu.: Modelirovanie fotosinteza i transpiratsii rasteniya v usloviyakh vodnogo defitsita. I. Obosnovanie i opisanie modeli. [Modelling plant photosynthesis and transpiration under water stress conditions. I. Theoretical basis and description of the model.] - In: UNGER, K., STÖCKER, G. (ed.): Biophysikalische Ökologie und Ökosystemforschung. Pp. 25-39. Akademie-Verlag, Berlin 1981. [In R, ab: E.]

8252 - BIKHELE, Z., MOLDAU, Kh., ROSS, Yu.: Modelirovanie fotosinteza i transpiratsii rasteniya v usloviyakh vodnogo defitsita. II. Rezul'taty chislennykh ékspe-rimentov. [Modelling plant photosynthesis and transpiration under water stress conditions. II. Results of numerical experiments.] - In: UNGER, K., STÖCKER, G. (ed.): Biophysikalische Ökologie und Ökosystemforschung. Pp. 41-60. Aka-demie-Verlag, Berlin 1981. [In R, ab: E.]

8253 - BÎNDIU, C., MIHALCIUC, V.: Modifications of the hydric state in green organs and stems after crown partial destruction by snow. - Rev. Roum. Biol., Sér. Biol. vég. 26: 11-18, 1981.

8254 - BIR, S.S., SATIJA, C.K., GOYAL, P., KAUR, S.: Stomatal patterns on some Asplenioid ferns from India. - Biol. Bull. Ind. 3: 24-35, 1981.

8255 - BIRYUKOVA, Z.P., KHARLAMOVA, N.V.: Geograficheskaya izmenchivost zasukho-ustoĭchivosti i zimostoĭkosti sosny obyknovennoĭ. [Geographic variability in drought resistance and winter hardiness in Scots pine.] - Ékologiya 1981 (4): 42-47, 1981. [In R.]

*8256 - BISSON, M.A., GUTKNECHT, J.: Osmotic regulation in algae. - In: SPANSWICK, R.M., LUCAS, W.J., DAINTY, J. (ed.): Plant Membrane Transport: Current Conceptual Issues. Pp. 131-142. Elsevier / North-Holland Biomedical Press, Amsterdam - New York - Oxford 1980.

*8257 - BISSON, M.A., KIRST, G.O.: A brackish water charophyte, *Lamprothamnium*: membrane PD and osmotic responses. - In: SPANSWICK, R.M., LUCAS, W.J., DAINTY, J. (ed.): Plant Membrane Transport: Current Conceptual Issues. Pp. 603-604. Elsevier / North-Holland Biomedical Press, Amsterdam - New York - Oxford 1980.

*8258 - BISSON, M.A., KIRST, G.O.: *Lamprothamnium*, a euryhaline charophyte. I. Osmotic relations and membrane potential at steady state. - J. exp. Bot. 31: 1223-1235, 1980.

*8259 - BISSON, M.A., KIRST, G.O.: *Lamprothamnium*, a euryhaline charophyte. II. Time course of turgor regulation. - J. exp. Bot. 31: 1237-1244, 1980.

*8260 - BJÖRKMAN, O.: The response of photosynthesis to temperature. - In: GRACE, J., FORD, E.D., JARVIS, P.G. (ed.): Plants and their Atmospheric Environment. Pp. 273-301. Blackwell Scientific Publication, Oxford - London - Edinburgh - Boston - Melbourne 1980.

8261 - BJÖRKMAN, O.: Responses to different quantum flux densities. - In: LANGE, O.L., NOBEL, P.S., OSMOND, C.B., ZIEGLER, H. (ed.): Physiological Plant Ecology I. Responses to the Physical Environment. Pp. 57-107. Springer-Verlag, Berlin - Heidelberg - New York 1981.

*8262 - BLACK, T.A., NNYAMAH, J.U.: A simple diffusion model of transpiration applied to a thinned Douglas-fir stand. - Ecology 59: 1221-1229, 1978.

*8263 - BLACK, T.A., TAN, C.S., NNYAMAH, J.U.: Transpiration rate of Douglas fir trees in thinned and unthinned stands. - Can. J. Soil Sci. 60: 625-631, 1980.

8264 - BLACQUIÈRE, T., LAMBERS, H.: Growth, photosynthesis and respiration in *Plantago coronopus* as affected by salinity. - Physiol. Plant. 51: 265-268, 1981.

8265 - BLAKE, T.J., REID, D.M.: Ethylene, water relations and tolerance to water-logging of three *Eucalyptus* species. - Aust. J. Plant Physiol. 8: 497-505, 1981.

8266 - BLAKER, N.S., MacDONALD, J.D.: Predisposing effects of soil moisture extremes on the susceptibility of rhododendron to *Phytophthora* root and crown rot. - Phytopathology 71: 831-834, 1981.

*8267 - BLIZZARD, W.E., BOYER, J.S.: Comparative resistance of the soil and the plant to water transport. - Plant Physiol. 66: 809-814, 1980.

*8268 - BLOEMEN, G.W.: Calculation of hydraulic conductivities of soils from texture and organic matter content. - Z. Pflanzenernähr. Bodenk. 143: 581-605, 1980.

*8269 - BLOEMEN, G.W.: Calculation of steady state capillary rise from the groundwater table in multi-layered soil profiles. - Z. Pflanzenernähr. Bodenk. 143: 701-719, 1980.

*8270 - BLOOM, A.J., MOONEY, H.A., BJÖRKMAN, O., BERRY, J.: Materials and methods for carbon dioxide and water exchange analysis. - Plant Cell Environ. 3: 371-375, 1980.

*8271 - BLUM, A., SINMENA, B., ZIV, O.: An evaluation of seed and seedling drought tolerance screening tests in wheat. - Euphytica 29: 727-736, 1980.

*8272 - **BLYUM, O.B., GIZBULINA, V.K.**: Katalazna aktyvnist' lyshaĭnykiv yak pokaznyk
ĭkh fiziologichnoĭ reaktsiĭ na vplyv vodnogo ta temperaturnogo faktoriv.
[The catalase activity of lichens as an indication of their physiological
response to the effect of water and temperature.] - Ukr. bot. Zh. 37(6):
49-54, 1980. [In Ukr, ab: R, E.]

8273 - **BOCZ, E., NAGY, J.**: A kukorica víz- és tápanyagellátásának optimalizálása
és hatása a termés tömegére. [Optimizing of water- and nutrient supply for
maize and effect on crop yield.] - Növénytermelés 30: 539-549, 1981. [In
Hung, ab: E.]

8274 - **BODSWORTH, S., BEWLEY, J.D.**: Osmotic priming of seeds of crop species with
polyethylene glycol as a means of enhancing early and synchronous germination
at cool temperatures. - Can. J. Bot. 59: 672-676, 1981.

*8275 - **BOĬKOV, S.**: Zavisimost mezhdu dobiva i vodata pri ogranichena vodoosigurenost
na lyutserna. [The relationship between yield and water consumption under
limited water supply of alfalfa.] - Rasteniev. Nauki 17 (7): 48-53, 1980.
[In Bulg, ab: E, R.]

8276 - **BOĬKOV, S.**: Evapotranspiratsiya i zavisimostta i s dobiva na semena ot
lyutserna. [Evapotranspiration and its relation with alfalfa seed yield.] -
Rasteniev. Nauki 18 (2): 81-88, 1981. [In Bulg, ab: E, R.]

8277 - **BOĬKOV, S.**: Mikroraĭonirane na polivniya rezhim na lyutsernata za furazh v"v
Vidinski okr"g. [Microregions of fodder alfalfa irrigation regime in the
Vidin district.] - Rasteniev. Nauki 18 (3): 81-87, 1981. [In Bulg, ab: E, R.]

8278 - **BOLTON, F.E.**: Optimizing the use of water and nitrogen through soil and crop
management. - Plant Soil 58: 231-247, 1981. Also in: MONTEITH, J., WEBB, C.
(ed.): Soil Water and Nitrogen in Mediterranean-type Environments. Development in Plant and Soil Sciences. Volume 1. Pp. 231-247, Martinus Nijhoff /
Dr. W. Junk Publishers, The Hague - Boston - London 1981.

*8279 - **BOLTON, J.K., BROWN, R.H.**: Photosynthesis of grass species differing in
carbon dioxide fixation pathways V. Response of *Panicum maximum*, *Panicum
milioides* and tall fescue (*Festuca arundinacea*) to nitrogen nutrition. -
Plant Physiol. 66: 97-100, 1980.

*8280 - **BORCHERT, R.**: Phenology and ecophysiology of tropical trees: *Erythrina
poeppigiana* O.F. Cook. - Ecology: 61: 1065-1074, 1980.

8281 - **BOROWITZKA, L.J.**: Solute accumulation and regulation of cell water activity.
- In: PALEG, L.G., ASPINALL, D. (ed.): The Physiology and Biochemistry of
Drought Resistance in Plants. Pp. 97-130. Academic Press, Sydney - New York -
London - Toronto - San Francisco 1981.

*8282 - **BOROWITZKA, L.J., DEMMERLE, S., MACKAY, M.A., NORTON, R.S.**: Carbon-13
nuclear magnetic resonance study of osmoregulation in a blue-green alga. -
Science 210: 650-651, 1980.

8283 - **BOTTINI, R., RACCA, R.W., ARGUELLO, J.A., COLLINO, D., TIZIO, R.**: Levels of
growth inhibitors in soybean (*Glycine max* L. Merrill) cv. Lee, in response
to drought, gibberellic acid and abscisic acid. - Fyton 41: 97-102, 1981.

8284 - **BRADFORD, K.J., YANG, S.F.**: Physiological responses of plants to waterlogging.
- HortScience 16: 25-30, 1981.

8285 - **BRAINERD, K.E., FUCHIGAMI, L.H., KWIATKOWSKI, S., CLARK, C.S.**: Leaf anatomy
and water stress of aseptically cultured 'Pixy' plum grown under different
environments. - HortScience 16: 173-175, 1981.

8286 - BRAKKE, T.W., KANEMASU, E.T., STEINER, J.L., ULABY, F.T., WILSON, E.: Micro-
       wave radar response to canopy moisture, leaf-area index, and dry weight of
       wheat, corn, and sorghum. - Remote Sensing Environ. 11: 207-220, 1981.

8287 - BRAMM, A.: Plant-water relations of rapeseed. - In: BUNTING, E.S. (ed.):
       Production and Utilization of Protein in Oilseed Crops. World Crops:
       Production, Utilization, and Description. Volume 5. Pp. 190-199. Martinus
       Nijhoff, The Hague - Boston - London 1981.

8288 - BRANDHAM, P.E., CUTLER, D.F.: Polyploidy, chromosome interchange and leaf
       surface anatomy as indicators of relationships within *Haworthia* section
       Coarctatae Baker (Liliaceae - Aloineae). - J. South Afr. Bot. 47: 507-546,
       1981.

8289 - BRAY, E.A., PARSONS, L.R.: Clonal variations in the water relations of red
       osier dogwood during cold acclimation. - Can. J. Plant Sci. 61: 351-363,
       1981.

8290 - BRENNAN, E., LEONE, I., HARKOV, R., RHOADS, A.: Austrian pine injury traced
       to ozone and sulfur dioxide pollution. - Plant Dis. 65: 363-364, 1981.

8291 - BRESSAN, R.A., HASEGAWA, P.M., HANDA, A.K.: Resistance of cultured higher
       plant cells to polyethylene glycol-induced water stress. - Plant Sci. Lett.
       21: 23-30, 1981.

8292 - BRETELER, H., GREENWOOD, D.J., PETTERSON, I., RUSSELL, J.S., SAUERBECK, D.,
       Van DORP, F., Van KEULEN, H., Van der MEER, H.G.: Soil-plant relations. -
       In: FRISSEL, M.J., Van VEEN, J.A. (ed.): Simulation of Nitrogen Behaviour
       of Soil-Plant Systems. Pp. 45-47. Pudoc, Wageningen 1981.

*8293 - BRISKE, D.D., WILSON, A.M.: Moisture and temperature requirements for
       adventitious root development in blue grama seedlings. - J. Range Manage.
       31: 174-178, 1978.

*8294 - BRISKE, D.D., WILSON, A.M.: Drought effects on adventitious root development
       in blue grama seedlings. - J. Range Manage. 33: 323-327, 1980.

8295 - BRISTOW, K.L., VAN ZYL, W.H., DE JAGER, J.M.: Measurement of leaf water
       potential using the J14 press. - J. exp. Bot. 32: 851-854, 1981.

8296 - BROSCHAT, T.K., DONSELMAN, H.M.: Effects of light intensity, air layering,
       and water stress on leaf diffusive resistance and incidence of leafspotting
       in *Ficus elastica*. - HortScience 16: 211-212, 1981.

*8297 - BROSE, E., FREIRE, J.R.J., MIELNICZUK, J.: Efeito da umidade e luminosidade
       sobre a atividade da nitrogenase nos nódulos de soja (*Glycine max* (L.)
       Merrill). [Effect of soil moisture and light on nitrogenase activity of no-
       dules of soybean (*Glycine max* (L.) Merrill).] - Agron. Sulriograndense 15:
       239-250, 1979. [In Port, ab: E.]

8298 - BROWN, D.H., SNELGAR, W.P., GREEN, T.G.A.: Effect of storage conditions on
       lichen respiration and desiccation sensitivity. - Ann. Bot. 48: 923-926,
       1981.

8299 - BROWN, P.W., TANNER, C.B.: Alfalfa water potential measurement: A comparison
       of the pressure chamber and leaf dew-point hygrometers. - Crop Sci. 21:
       240-244, 1981.

8300 - BROWN, S.: A comparison of the structure, primary productivity, and transpi-
       ration of cypress ecosystems in Florida. - Ecol. Monogr. 51: 403-427, 1981.

*8301 - BÜCHNER, K.-H., BENTRUP, F.-W., ZIMMERMANN, U.: Pressure-probe measurements of water transport parameters in suspension-cultured cells of *Chenopodium rubrum* L. - In: SPANSWICK, R.M., LUCAS, W.J., DAINTY, J. (ed.): Plant Membrane Transport: Current Conceptual Issue. Pp. 473-474. Elsevier / North-Holland Biomedical Press, Amsterdam - New York - Oxford 1980.

8302 - BÜCHNER, K.-H., ZIMMERMANN, U., BENTRUP, F.-W.: Turgor pressure and water transport properties of suspension-cultured cells of *Chenopodium rubrum* L. - Planta 151: 95-102, 1981.

8303 - BUCKS, D.A., ERIE, L.J., FRENCH, O.F., NAKAYAMA, F.S., PEW, W.D.: Subsurface trickle irrigation management with multiple cropping. - Trans. ASAE 24: 1482-1489, 1981.

8304 - BUDKEVICH, T.A.: Vodnyĭ rezhim klevera krasnogo, lyutserny i timofeevki v svyazi s ikh vzaimodeĭstviem v travosmesyakh. [Water regime of red clover, alfalfa and timothy grass in view of their relationships in grass mixtures.] - Vestsi Akad. Navuk Belarus. SSR, Ser. biyal. Navuk 1981 (3): 107, 1981. [In R.]

8305 - BUHTZ, E., FELGNER, G., BAHN, E., BÄTZ, G.: Ergebnisse unterschiedlicher Einsatzformen der Beregnung zu Zuckerrüben. - Arch. Acker- Pflanzenbau Bodenk. 25: 117-124, 1981.

8306 - BUNCE, J.A.: Relationships between maximum photosynthetic rates and photosynthetic tolerance of low leaf water potentials. - Can. J. Bot. 59: 769-774, 1981.

8307 - BUNCE, J.A.: Comparative responses of leaf conductance to humidity in single attached leaves. - J. exp. Bot. 32: 629-634, 1981.

8308 - BURKINA, Z.S., GUSEINOVA, G.M.: Application of $H_2^{18}O$ for studying water exchange in maize roots in relation to energy supply and membrane state. - Studia biophys. 85: 21-22, 1981.

8309 - BYRNE, M.C., NELSON, C.J., RANDALL, D.D.: Ploidy effects on anatomy and gas exchange of tall fescue leaves. - Plant Physiol. 68: 891-893, 1981.

*8310 - CAIN, D.W., OLIEN, C.R., ANDERSEN, R.L.: Comparative freezing patterns of bark and xylem of "Siberian C" and "Redhaven" peach twigs. - Cryobiology 17: 495-499, 1980.

8311 - CALDWELL, M.M., RICHARDS, J.H., JOHNSON, D.A., NOWAK, R.S., DZUREC, R.S.: Coping with herbivory: Photosynthetic capacity and resource allocation in two semiarid *Agropyron* bunchgrasses. - Oecologia 50: 14-24, 1981.

8312 - CALERO, E., WEST, S.H., HINSON, K.: Water absorption of soybean seeds and associated causal factors. - Crop Sci. 21: 926-933, 1981.

*8313 - CALIANDRO, A., TARANTINO, E., RUBINO, P.: Consumi idrici della barbabietola da zucchero a semina primaverile in ambiente dell'Italia meridionale. [Water consumption by spring sown sugar beet in southern Italy.] - Riv. Agron. 14: 178-193, 1980. [In Ital, ab: E.]

8314 - CAMPBELL, C.A., DAVIDSON, H.R., WINKLEMAN, G.E.: Effect of nitrogen, temperature, growth stage and duration of moisture stress on yield components and protein content of Manitou spring wheat. - Can. J. Plant Sci. 61: 549-563, 1981.

8315 - CAMPBELL, G.S.: Fundamentals of radiation and temperature relations. - In: LANGE, O.L., NOBEL, P.S., OSMOND, C.B., ZIEGLER, H. (ed.): Physiological Plant Ecology I. Responses to the Physical Environment. Pp. 11-40. Springer -Verlag, Berlin - Heidelberg - New York 1981.

8316 - CAMPBELL, G.S., HARRIS, G.A.: Modeling soil-water-plant-atmosphere systems of deserts. - In: EVANS, D.D., THAMES, J.L. (ed.): Water in Desert Ecosystems. Pp. 75-91. Dowden, Hutchinson & Ross, Inc., Stroudsburg 1981.

8317 - CAMPBELL, R.B., REICOSKY, D.C., DOTY, C.W.: Net radiation within a canopy of sweet corn during drought. - Agr. Meteorol. 23: 143-150, 1981.

*8318 - CAMPBELL, W.F., WAGENET, R.J., BAMATRAF, A.M., TURNER, D.L.: Path coefficient analysis of correlation between stress and barley yield components. - Agron. J. 72: 1012-1016, 1980.

8319 - CARBONNIER, J., GIRAUD, M., HUBAC, C., MOLHO, D., VALLA, A.: Activité anti-transpirante d'analogues de l'acide abscissique. - Physiol. Plant. 51: 1-6, 1981.

8320 - CARLSON, J.R.,Jr., DITTERLINE, R.L., MARTIN, J.M., LUND, R.E.: Sampling stomatal density in alfalfa. - Crop Sci. 21: 467-469, 1981.

*8321 - CARPENA, O., PÉREZ MELIÁN, G., LUQUE, A.: Absorción de agua e iones en el cultivo de pepinos. I. Consumos totales. [Water and ions absorption by cucumber plants in hydroponics. I. Total uptake.] - Rev. Agroquím. Tecnol. Aliment. 18: 236-244, 1978. [In Span, ab: E.]

8322 - CARPITA, N.C., NABORS, M.W.: Growth physics and water relations of red-light--induced germination in lettuce seeds. V. Promotion of elongation in the embryonic axes by gibberellins and phytochrome. - Planta 152: 131-136, 1981.

*8323 - CARR, D.J., CARR, S.G.M.: Correlation between sizes of stomata and of oil glands in two related *Eucalyptus* species. - Aust. J. Bot. 28: 551-554, 1980.

*8324 - CARTER, M.R.: Effects of sulphate and chloride soil salinity on growth and needle composition of Siberian larch. - Can. J. Plant Sci. 60: 903-910, 1980.

8325 - CARY, J.W.: Calculation of CO_2 gas phase diffusion in leaves and its relation to stomatal resistance. - Photosynthesis Res. 2: 185-194, 1981.

8326 - CATHEY, G.W., ELMORE, C.D., McMICHAEL, B.L.: Some physiological responses of cotton leaves to foliar applications of potassium-3,4-dichloroisothiazole-5-carboxylate and S,S,S-tributyl phosphorotrithioate. - Physiol. Plant. 51: 140-144, 1981.

8327 - ČATSKÝ, J., TICHÁ, I.: Transport a cesty oxidu uhličitého ve fotosyntetizujícím listu. [Transport and pathways of carbon dioxide in a photosynthesizing leaf.] - Biol. Listy (Praha) 46: 1-26, 1981. [In Czech, ab: E.]

*8328 - CATTESON, A.M.: The vascular cambium. - In: LITTLE, C.H.A. (ed.): Control of Shoot Growth in Trees. Pp. 12-40, Maritimes Forest Research Centre, Fredericton 1980.

*8329 - CERDA, A., BINGHAM, F.T., LABANAUSKAS, C.K.: Blossom-end rot of tomato fruit as influenced by osmotic potential and phosphorous concentrations of nutrient solution media. - J. Amer. Soc. hort. Sci. 104: 236-239, 1979.

*8330 - ČERMÁK, J., KUČERA, J.: Sezónní průběh transpiračního proudu a spotřeba vody u dubu (*Quercus robur* L.) v lužním lese. [Seasonal course of sap flow rate and water consumption of full-grown oak (*Quercus robur* L.) in floodplain forest.] - In: Zborník Referátov 3. Zjazdu Slovenskej Botanickej Spoločnosti Zvolen 1980. Pp. 233-238. Slovenská Botanická Spoločnost, Bratislava 1980. [In Czech, ab: R, E.]

8331 - ČERMÁK, J., KUČERA, J.: The compensation of natural temperature gradient at the measuring point during sap flow rate determination in trees. - Biol. Plant. 23: 469-471, 1981.

*8332 - CERNUSCA, A., SEEBER, M.C.: Canopy structure, microclimate and the energy budget in different alpine plant communities. - In: GRACE, J., FORD, E.D., JARVIS, P.G. (ed.): Plants and their Atmospheric Environment. Pp. 75-81. Blackwell Scientific Publications, Oxford - London - Edinburgh - Boston - Melbourne 1980.

8333 - CEULEMANS, R., GABRIËLS, R., IMPENS, I.: Influence of nutritional status on some ecophysiological, morphological and architectural characteristics of *Azalea*. - Gartenbauwissenschaft 46: 206-208, 1981.

8334 - CEULEMANS, R., IMPENS, I.: Le modèle d'analogie électrique en tant que moyen d'étude de l'assimilation et de la transpiration des plantes. Application aux échanges gazeux au niveau de la feuille du peuplier et leur liaison à la croissance et à la productivité. - Rev. Agr. 34: 1581-1593, 1981.

*8335 - CHAHAL, R.S., SINGH, B.P., GUPTA, A.P., KALA, R.: Production potential of various crops under different levels of fertilizers and irrigation. - Ind. J. Agron. 25: 358-361, 1980.

8336 - CHALMERS, D.J., MITCHELL, P.D., HEEK, L., van: Control of peach tree growth and productivity by regulated water supply, tree density, and summer pruning. - J. Amer. Soc. hort. Sci. 106: 307-312, 1981.

*8337 - CHAMBERLAIN, A.C., LITTLE, P.: Transport and capture of particles by vegetation. - In: GRACE, J., FORD, E.D., JARVIS, P.G. (ed.): Plants and their Atmospheric Environment. Pp. 147-173. Blackwell Scientific Publications, Oxford - London - Edinburgh - Boston - Melbourne 1980.

8338 - CHAN, K.-Y., WONG, K.H., NG, S.L.: Effects of polyethylene glycol on growth and cadmium accumulation of *Chlorella salina* CU-1. - Chemosphere 10: 985-991, 1981.

8339 - CHANEY, W.R.: Sources of water. - In: KOZLOWSKI, T.T. (ed.): Water Deficits and Plant Growth. Volume VI. Woody Plant Communities. Pp. 1-47. Academic Press, New York - San Francisco - London 1981.

8340 - CHAPMAN, D.C., PARKER, R.L.: A theoretical analysis of the diffusion porometer: Steady diffusion through two finite cylinders of different radii. - Agr. Meteorol. 23: 9-20, 1981.

*8341 - CHAPMAN, K.R., CREW, P.: Influence of five watering frequencies regulated by trickle irrigation, on the growth and cropping of apple trees in Queensland. - Queensland J. agr. anim. Sci. 35: 105-119, 1978.

*8342 - CHAVAN, P.D., KARADGE, B.A.: Influence of sodium chloride and sodium sulfate salinities on photosynthetic carbon assimilation in peanut. - Plant Soil 56: 201-207, 1980.

*8343 - CHEN, Y., ZAHAVI, E., BARAK, P., UMIEL, N.: Effects of salinity stresses on tobacco. I. The growth of *Nicotiana tabacum* callus cultures under seawater, NaCl, and mannitol stresses. - Z. Pflanzenphysiol. 98: 141-153, 1980.

*8344 - CHERNOVOL, A.E.: Urozhaĭ ogurtsov v zavisimosti ot vremeni polivov. [Cucumber yield in dependence on time of irrigation.] - Nauch.-tekh. Byull. (Khar'kov) 1979(10): 8-14, 1979. [In R.]

8345 - CHETAL, S., WAGLE, D.S., NAINAWATEE, H.S.: Glycolipid changes in wheat and barley chloroplast under water stress. - Plant Sci. Lett. 20: 225-230, 1981.

8346 - CHHABRA, M.L., DHINGRA, H.R., YADAVA, T.P.: Screening of Indian mustard (*Brassica juncea* L. Czern and Coss) varieties for drought resistance. - Ind. J. Plant Physiol. 24: 8-11, 1981.

8347 - **CHOUDHURI, G.N., VARSHNEY, S.P.:** Profile moisture and primary productivity of *Cynodon dactylon* (Pers) in a salt-affected habitat. - Comp. Physiol. Ecol. 6: 53-56, 1981.

8348 - **CHRISTEN, A.A.:** Verticillium wilt in alfalfa. - Plant Dis. 65: 319-321, 1981.

8349 - **CHRISTENSEN, N.W., TAYLOR, R.G., JACKSON, T.L., MITCHELL, B.L.:** Chloride effects on water potentials and yield of winter wheat infected with take-all root rot. - Agron. J. 73: 1053-1058, 1981.

*8350 - **CHUNG, I., STADELMANN, E.J., KWON, Y.M.:** Effect of BASF 13-338 on drought tolerance, osmotic value and passive permeability of cells from *Hordeum vulgare* L., cv. Baecdong seedlings. - Plant Physiol. 65 (Suppl.): 155, 1980.

8351 - **CLARK, R.B.:** Effect of light and water stress on mineral element composition of plants. - J. Plant Nutr. 3: 853-885, 1981.

*8352 - **CLARK, S.B., LETEY, J.,Jr., LUNT, O.R., WALLACE, A., KLEINKOPF, G.E., ROMNEY, E.M.:** Transpiration and CO_2 fixation of selected desert shrubs as related to soil-water potential. - Great Basin Naturalist Memoirs 1980 (4 - Soil-Plant-Animal Relationships Bearing on Revegetation and Land Reclamation in Nevada Deserts): 110-116, 1980.

8353 - **CLARKE, J.M.:** Effect of harvest time and drying method on quality and grade of irrigated soft white spring wheat. - Can. J. Plant Sci. 61: 803-810, 1981.

8354 - **CLARKSON, D.T.:** Nutrient interception and transport by root systems. - In: JOHNSON, C.B. (ed.): Physiological Processes Limiting Plant Productivity. Pp. 307-330. Butterworths, London - Boston - Sydney - Wellington - Durban - Toronto 1981.

8355 - **CLAUS, S.:** Biophysikalische Systemanalyse der Ertragsbildung von Getreide. - In: UNGER, K., STÖCKER, G. (ed.): Biophysikalische Ökologie und Ökosystem-forschung. Pp. 61-67. Akademie-Verlag, Berlin 1981.

8356 - **COCHRANE, T.T., JONES, P.G.:** Savannas, forests and wet season potential evapotranspiration in tropical South America. - Trop. Agr. 58: 185-190, 1981.

8357 - **COCKBURN, W.:** The evolutionary relationship between stomatal mechanism, crassulacean acid metabolism and C_4 photosynthesis. - Plant Cell Environ. 4: 417-418, 1981.

*8358 - **COHEN, M.:** Hydraulic conductivity of wood of trees with citrus blight. - Proc. Florida State hort. Soc. 92: 68-70, 1979.

8359 - **COHEN, Y., BLACK, T.A., KELLIHER, F.M.:** Determination of sap flow in Douglas-fir trees using the heat pulse technique. - In: Preprint Volume of Extended Abstracts: 15th Conference on Agriculture and Forest Meteorology and Fifth Conference on Biometeorology. Pp. 166-169. American Meteorological Society, Boston 1981.

*8360 - **COLTON, C.E., EINHELLIG, F.A.:** Allelopathic mechanisms of velvetleaf (*Abutilon theophrasti* Medic., Malvaceae) on soybean. - Amer. J. Bot. 67: 1407-1413, 1980.

8361 - **COMBE, L.:** Effet d'une fumure carbonée sur les capacités d'assimilation nette du radis selon l'éclairement de culture. - Agronomie 1: 93-98, 1981.

8362 - **COMBE, L.:** Effet du gaz carbonique et de la culture en climat artificiel sur la croissance et le rendement d'un blé d'hiver. - Agronomie 1: 177-185, 1981.

8363 - **COMBER, R.**: The effect of water supply on the growth of Izmir, Palotina and Amarelinho tobaccos. - Beitr. Tabakforsch. Int. 11: 99-105, 1981.

8364 - **COMERFORD, N.B., LEAF, A.L.**: The mean accuracy and the precision of using stem discs moisture contents to estimate stem dry weight of *Pinus resinosa* Ait. - Forest Ecol. Manage. 3: 329-334, 1980/1981.

8365 - **CONARD, S.G., RADOSEVICH, S.R.**: Photosynthesis, xylem pressure potential, and leaf conductance of three montane chaparral species in California. - Forest Sci. 27: 627-639, 1981.

8366 - **CONNER, W.H., GOSSELINK, J.G., PARRONDO, R.T.**: Comparison of the vegetation of three Louisiana swamp sites with different flooding regimes. - Amer. J. Bot. 68: 320-331, 1981.

8367 - **CONNOR, D.J., COCK, J.H., PARRA, G.E.**: Response of cassava to water shortage. I. Growth and yield. - Field Crops Res. 4: 181-200, 1981.

8368 - **CONSTABLE, G.A., HEARN, A.B.**: Irrigation for crops in a sub-humid environment. VI. Effect of irrigation and nitrogen fertilizer on growth, yield and quality of cotton. - Irrig. Sci. 3: 17-28, 1981.

*8369 - **COOK, M.**: Peanut leaf wettability and susceptibility to infection by *Puccinia arachidis*. - Phytopathology 70: 826-830, 1980.

8370 - **COOK, M.**: A quick technique for determining the wettability of leaves of *Arachis hypogaea* and certain other species. - Peanut Sci. 8: 57-60, 1981.

8371 - **COOPER, J.P.**: Physiological constraints to varietal improvement. - Phil. Trans. roy. Soc. London B 292: 431-440, 1981.

8372 - **COOPER, K.M., TINKER, P.B.**: Translocation and transfer of nutrients in vesicular-arbuscular mycorrhizas. IV. Effect of environmental variables on movement of phosphorus. - New Phytol. 88: 327-340, 1981.

8373 - **COPONY, W.**: Einfluss von Klima und Boden auf die schlagbezogene Optimierung der Düngung von Kartoffeln. - In: UNGER, K., STÖCKER, G. (ed.): Biophysikalische Ökologie und Ökosystemforschung. Pp. 163-169. Akademie-Verlag, Berlin 1981.

8374 - **CORNILLON, P., DAUPLE, P.**: Influence of irrigation rhythm and water supply on growth, water status and yield of egg-plant (*Solanum melongena* L.). - Plant Soil 59: 365-379, 1981.

8375 - **CORNISH, P.S.**: Resistance to water flow in the intracoleoptile internode of wheat. - Plant Soil 59: 119-125, 1981.

8376 - **COSGROVE, D., STEUDLE, E.**: Water relations of growing pea epicotyl segments. - Planta 153: 343-350, 1981.

8377 - **COSGROVE, D.J.**: Analysis of the dynamic and steady-state responses of growth rate and turgor pressure to changes in cell parameters. - Plant Physiol. 68: 1439-1446, 1981.

8378 - **COTHREN, J.T., STUTTE, C.A., RUTLEDGE, S.R.**: Soybean yields as influenced by tributyl[(5-chloro-2-thienyl)methyl]phosphonium chloride and different moisture regimes. - Phyton 40: 117-125, 1981.

8379 - **COUDRET, A.**: Action du NaCl sur les contraintes et les relations hydriques dans les parties aériennes de *Plantago maritima* L. et *Plantago lanceolata* L. - Acta Oecol. - Oecol. Plant. 2: 111-120, 1981.

*8380 - COUGHLAN, S.J., WYN JONES, R.G.: Some responses of *Spinacia oleracea* to salt stress. - J. exp. Bot. 31: 883-893, 1980.

8381 - COUTTS, M.P.: Effects of waterlogging on water relations of actively-growing and dormant Sitka spruce seedlings. - Ann. Bot. 47: 747-753, 1981.

8382 - COUTTS, M.P.: Effects of root or shoot exposure before planting on the water relations, growth, and survival of Sitka spruce. - Can. J. Forest Res. 11: 703-709, 1981.

8383 - COUTTS, M.P.: Leaf water potential and control of water loss in droughted Sitka spruce seedlings. - J. exp. Bot. 32: 1193-1201, 1981.

8384 - COX, E.F., DEARMAN, A.S.: The effect of trickle irrigation, misting and row position on the incidence of tipburn of field lettuce. - Scientia Hort. 15: 101-106, 1981.

8385 - COYNE, P.I., BINGHAM, G.E.: Comparative ozone dose response of gas exchange in a ponderosa pine stand exposed to long-term fumigations. - J. Air Pollut. Control Assoc. 31: 38-41, 1981.

*8386 - CRAM, J.: The higher plant as a whole. - In: SPANSWICK, R.M., LUCAS, W.J., DAINTY, J. (ed.): Plant Membrane Transport: Current Conceptual Issues. Pp. 3-13. Elsevier / North-Holland Biomedical Press, Amsterdam - New York - Oxford 1980.

8387 - CREATURA, P.J., SAFIR, G.R., SCHEFFER, R.P., SHARKEY, T.D.: Effects of *Cephalosporium gramineum* and a toxic metabolite on stomatal conductance of wheat. - Physiol. Plant Pathol. 19: 313-323, 1981.

8388 - CRIPPS, J.E.L.: Biennial patterns in apple tree growth and cropping as related to irrigation and thinning. - J. hort. Sci. 56: 161-168, 1981.

8389 - CROSBIE, T.M., PEARCE, R.B., MOCK, J.J.: Selection for high CO_2 exchange rate among inbred lines of maize. - Crop Sci. 21: 629-631, 1981.

*8390 - CRUIZIAT, P.: L'eau et les cultures. - In: Encyclopédie des Techniques Agricoles. Volume 1165. Pp. 1-30. Association pour le Développement et la Vulgarisation des Techniques Agricoles, Paris 1980.

*8391 - CSERESNYES, Z.: Studies on the duration of dormancy and methods of determining the germination of dormant seeds of *Helianthus annuus*. - Seed Sci. Technol. 7: 179-188, 1979.

*8392 - CSERESNYES, Z.: The germination of *Helianthus annuus* seeds under optimum laboratory conditions. - Seed Sci. Technol. 7: 319-328, 1979.

*8393 - CSIZINSKY, A.A.: The importance of irrigation frequency and fertilizer placement in growing vegetables with drip irrigation. - Proc. Florida State hort. Soc. 92: 76-80, 1979.

*8394 - CSIZINSZKY, A.A.: Yield and water use of vegetable crops with seepage and drip irrigation systems. - Florida Sci. 43: 285-292, 1980.

*8395 - CURE, J., PATTERSON, R.P., RAPER, C.D.: Effects of moderate water stress on assimilate partitioning in soybean plants with normal and with photoperiodically reduced reproductive load. - Plant Physiol. 65 (Suppl.): 7, 1980.

*8396 - CURRAN, P.M.T.: Vegetative growth of terrestrial and marine fungi in response to salinity. - Nowa Hedwigia 32: 285-292, 1980.

*8397 - CUTLER, D.F., HARTMANN, H.: Scanning Electron Microscope Studies of the Leaf Epidermis in Some Succulents (Tropische Subtropische Pflanzenwelt 28). - Akademie der Wissenschaften und der Literatur, Mainz 1979.

*8398 - CUTLER, J.M., SHAHAN, K.W., STEPONKUS, P.L.: Characterization of internal
         water relations of rice by a pressure-volume method. - Crop Sci. 19: 681-685,
         1979.

*8399 - CUTLER, J.M., STEPONKUS, P.L., WACH, M.J., SHAHAN, K.W.: Dynamic aspects and
         enhancement of leaf elongation in rice. - Plant Physiol. 66: 147-152, 1980.

 8400 - DAIE, J., CAMPBELL, W.F.: Response of tomato plants to stressful temperatures.
         Increase in abscisic acid concentrations. - Plant Physiol. 67: 26-29, 1981.

 8401 - DAINTY, J., KLEINOVÁ, M., JANÁČEK, K.: The movement of water across the plant
         root. - Plant Soil 63: 11-14, 1981.

*8402 - DAIYA, K.S., SHARMA, H.K., CHAWAN, D.D., SEN, D.N.: Effect of salt solutions
         of different osmotic potential on seed germination and seedlings growth in
         some Cassia spp. - Folia Geobot. Phytotaxon. 15: 149-154, 1980.

*8403 - DA MOTA, F.S.: Metodologia para caracterizaçao da seca agronômica no Brasil.
         [Methodology for the characterization of agricultural drought in Brasil.] -
         Interciencia 4: 344-349, 1979. [In Port, ab: E.]

 8404 - DA MOTA, F.S.: Índice de seca para soja. Contribuição para um modelo de
         previsão do rendimento da soja no Rio Grande do Sul. [A drought index for
         soybeans. A contribution to a soybean yield prediction model for Rio Grande
         do Sul.] - Pesq. agropec. Brasil. 16: 371-383, 1981. [In Port, ab: E.]

 8405 - DARBYSHIRE, B., ALLAWAY, W.G.: Soluble carbohydrates in leaf epidermis of
         Allium cepa: A potential role in stomatal function? - Plant Sci. Lett. 22:
         141-145, 1981.

 8406 - DA SILVA, J.E., RESCK, D.V.S.: Respostas fisiológicas da soja ao déficit
         hídrico em dois solos de cerrado. [Physiological response of soybean under
         water stress in two "cerrado" soils.] - Pesq. agropec. Brasil. 16: 669-675,
         1981. [In Port, ab: E.]

 8407 - DA SILVA, P.R.F., STUTTE, C.A.: Nitrogen loss in conjunction with transpi-
         ration from rice leaves as influenced by growth stage, leaf position, and
         N supply. - Agron. J. 73: 38-42, 1981.

 8408 - DA SILVEIRA, P.M., GUIMARÃES, C.M., STONE, L.F., KLUTHCOUSKI, J.: Avaliação
         de cultivares de feijão para resistência à seca baseada em dias se estresse
         de água no solo. [Evaluation of bean cultivars to drought resistance based
         on water stress days in soil.]- Pesq. agropec. Brasil. 16: 693-699, 1981.
         [In Port, ab: E.]

*8409 - DAULAY, H.S., SINGH, R.P.: Optimum utilization of limited water resources
         of arid lands for efficient crop production. - Ann. arid Zone 19: 203-213,
         1980.

*8410 - DAVENPORT, D.C., HAGAN, R.M.: A conceptual framework for evaluating agri-
         cultural water conservation potentials. - Water Resour. Bull. 16: 717-723,
         1980.

 8411 - DAVIDS, J.A., DAVIS, D.D., PENNYPACKER, S.P.: The influence of soil moisture
         on macroscopic sulfur dioxide injury to pinto bean foliage. - Phytopathology
         71: 1208-1212, 1981.

 8412 - DAVIES, F.S., BUCHANAN, D.W., ANDERSON, J.A.: Water stress and cold hardiness
         in field-grown citrus. - J. Amer. Soc. hort. Sci. 106: 197-200, 1981.

8413 - **DAVIES, W.J., WILSON, J.A., SHARP, R.E., OSONUBI, O.:** Control of stomatal behaviour in water-stressed plants. - In: JARVIS, P.G., MANSFIELD, T.A. (ed.): Stomatal Physiology. Pp. 163-185. Cambridge University Press, Cambridge - London - New York - New Rochelle - Melbourne - Sydney 1981.

8414 - **DAVIS, T.D., POTTER, J.R.:** Current photosynthate as a limiting factor in adventitious root formation on leafy pea cuttings. - J. Amer. Soc. hort. Sci. 106: 278-282, 1981.

8415 - **DAWES, C.J., McINTOSH, R.P.:** The effect of organic material and inorganic ions on the photosynthetic rate of the red alga *Bostrychia binderi* from a Florida estuary. - Mar. Biol. 64: 213-218, 1981.

8416 - **DAY, W.:** Water stress and crop growth. - In: JOHNSON, C.B. (ed.): Physiological Processes Limiting Plant Productivity. Pp. 199-215. Butterworths, London - Boston - Sydney - Wellington - Durban - Toronto 1981.

8417 - **DAY, W., LAWLOR, D.W., LEGG, B.J.:** The effects of drought on barley: soil and plant water relations. - J. agr. Sci. 96: 61-77, 1981.

8418 - **DEBERGH, P., HARBAOUI, Y., LEMEUR, R.:** Mass propagation of globe artichoke (*Cynara scolymus*): Evaluation of different hypotheses to overcome vitrification with special reference to water potential. - Physiol. Plant. 53: 181-187, 1981.

*8419 - **De GROOTH, B.G., Van GORKOM, H.J., MEIBURG, R.F.:** Generation of the 518 nm absorbance change in chloroplasts by an externally applied electrical field. - FEBS Lett. 113: 21-24, 1980.

*8420 - **DEHGAN, B.:** Application of epidermal morphology to taxonomic delimitations in the genus *Jatropha* L. (Euphorbiaceae). - Bot. J. Linn. Soc. 80: 257-278, 1980.

8421 - **DEJAEGERE, R., NEIRINCKX, Ł., STASSART, J.M., DELEGHER, V.:** Mechanism of ion uptake across barley roots. - Plant Soil 63: 19-24, 1981. Also in: BROUWER, R., GASPARIKOVÁ, O., KOLEK, J., LAUGHMAN, B.C. (ed.): Structure and Function of Plant Roots. Pp. 173-178. Martinus Nijhoff / Dr. W.Junk Publishers, The Hague - Boston - London 1981.

8422 - **De JONG, T.M., DRAKE, B.G.:** Seasonal patterns of plant and soil water potential on an irregularly-flooded tidal marsh. - Aquat. Bot. 11: 1-9, 1981.

8423 - **De JONG, T.M., PHILLIPS, D.A.:** Nitrogen stress and apparent photosynthesis in symbiotically grown *Pisum sativum* L. - Plant Physiol. 68: 309-313, 1981.

*8424 - **DE LA GUARDIA, M.D., BENLLOCH, M.:** Effects of potassium and gibberellic acid on stem growth of whole sunflower plants. - Physiol. Plant. 49: 443-448, 1980.

8425 - **DE MICHELIS, M.I., PUGLIARELLO, M.C., RASI-CALDOGNO, F., DE VECCHI, L.:** Osmotic behaviour and permeability properties of vesicles in microsomal preparations from pea internodes. - J. exp. Bot. 32: 293-302, 1981.

*8426 - **DE MICHELIS, M.I., RASI-CALDOGNO, F., PUGLIARELLO, M.C.:** Osmotic behaviour of vesicles in microsomal preparations from pea internodes. - In: SPANSWICK, R.M., LUCAS, W.J., DAINTY, J. (ed.): Plant Membrane Transport: Current Conceptual Issues. Pp. 519-520. Elsevier / North-Holland Biomedical Press, Amsterdam - New York - Oxford 1980.

*8427 - **DENNETT, M.D., ELTSON, J., DIEGO, Q.R.:** Weather and yields of tobacco, sugar beet and wheat in Europe. - Agr. Meteorol. 21: 249-263, 1980.

8428 - DESCHÊNES, J.-M., DUBUC, J.-P.: Effets de l'humidité du sol, des dates de
semis et des mauvaises herbes sur le rendement de céréales. - Can. J. Plant
Sci. 61: 851-857, 1981.

8429 - DE STIGTER, H.C.M.: Effects of glucose with 8-hydroxyquinoline sulfate or
aluminium sulfate on the water balance of cut "Sonia" roses. - Z. Pflanzenphysiol. 101: 95-105, 1981.

✶8430 - DEV CHOUDHURY, M.N., BAJAJ, K.L.: Biochemical changes during withering of
tea shoots. - Two Bud 27: 13-16, 1980.

8431 - DHINDSA, R.S., MATOWE, W.: Drought tolerance in two mosses: correlated with
enzymatic defence against lipid peroxidation. - J. exp. Bot. 32: 79-91,
1981.

8432 - DHINGRA, O.D., CHAGAS, D.: Effect of soil temperature, moisture, and nitrogen on competitive saprophytic ability of *Macrophomina phaseolina*. - Trans.
Brit. mycol. Soc. 77: 15-20, 1981.

8433 - DICKENSON, S., WHEELER, B.E.J.: Effects of temperature, and water stress in
sycamore, on growth of *Cryptostroma corticale*. - Trans. Brit. mycol. Soc.
76: 181-185, 1981.

✶8434 - DICKSON, D.M., WYN JONES, R.G., DAVENPORT, J.: Steady state osmotic adaptation in *Ulva lactuca*. - Planta 150: 158-165, 1980.

8435 - DIDDEN-ZOPFY, B.: Tagesgänge von Grössen des Wasserhaushaltes bei *Lolium
perenne* L. und *Arrhenatherum elatius* (L.) M. unter verschiedenen Umwelt-
bedigungen. - Acta Oecol. - Oecol. Plant. 2: 291-304, 1981.

✶8436 - DIEFFENBACH, H., KRAMER, D., LÜTTGE, U.: Release of guttation fluid from
passive hydathodes of intact barley plants. I. Structural and cytological
aspects. - Ann. Bot. 45: 397-401, 1980.

✶8437 - DIEFFENBACH, H., LÜTTGE, U., PITMAN, M.G.: Release of guttation fluid from
passive hydathodes of intact barley plants. II. The effects of abscisic
acid and cytokinins. - Ann. Bot. 703-712, 1980.

✶8438 - DIMITROVA-CHERVENKOVA, Z.: Efektivnost na proizvodstvo na tsarevitsa za
z"rno pri narushen rezhim na napoyavane. [Economic effectiveness in the
production of maize for grain in case of disturbed irrigation regime.] -
Rasteniev. Nauki 17(6): 92-96, 1980. [In Bulg, ab: E.]

✶8439 - DIRKS, V.A., BOLTON, E.F.: Regression analyses of grain yield of corn, level
of leaf N P K and soil conditions in a long-term rotation experiment on
Brookston clay. - Can. J. Soil Sci. 60: 599-612, 1980.

8440 - DIRKS, V.A., BOLTON, E.F.: Climatic factors contributing to year-to-year
variation in grain yield of corn on Brookston clay. - Can. J. Plant Sci. 61:
293-305, 1981.

8441 - DONKIN, M.E., MARTIN, E.S.: *In vivo* microspectrophotometry of guard cell
chloroplasts. - Z. Pflanzenphysiol. 102: 267-271, 1981.

8442 - DONKIN, M.E., MARTIN, E.S.: Blue light absorption by guard cells of *Commelina communis* and *Allium cepa*. - Z. Pflanzenphysiol. 102: 345-352, 1981.

✶8443 - DOSSKEY, M.G., BALLARD, T.M.: Resistance to water uptake by Douglas-fir
seedlings in soils of different texture. - Can. J. Forest Res. 10: 530-534,
1980.

8444 - **DOUGHERTY, C.T.**: Effect of soil moisture, solar radiation, and dew on wheat ear water potentials. - N. Zeal. J. agr. Res. 24: 139-141, 1981.

8445 - **DOUGHERTY, P.M., HINCKLEY, T.M.**: The influence of a severe drought on net photosynthesis of white oak (*Quercus alba*). - Can. J. Bot. 59: 335-341, 1981.

* 8446 - **DOUGHERTY, P.M., MORIKAWA, Y.**: Influence of elevation on the growth, photosynthesis and water relations of *Pseudotsuga menziesii*. - Plant Physiol. 65 (Suppl.): 155, 1980.

8447 - **DOUGLAS, T.J., PALEG, L.G.**: Lipid composition of *Zea mays* seedlings and water stress-induced changes. - J. exp. Bot. 32: 499-508, 1981.

8448 - **DOWNTON, W.J.S.**: Water relations of laticifers in *Nerium oleander*. - Aust. J. Plant Physiol. 8: 329-334, 1981.

8449 - **DOWNTON, W.J.S., LOVEYS, B.R.**: Abscisic acid content and osmotic relations of salt-stressed grapevine leaves. - Aust. J. Plant Physiol. 8: 443-452, 1981.

8450 - **DRAKE, S.R., PROEBSTING, E.L.,Jr., MAHAN, M.O., THOMPSON, J.B.**: Influence of trickle and sprinkle irrigation on 'Golden Delicious' apple quality. - J. Amer. Soc. hort. Sci. 106: 255-258, 1981.

*8451 - **DUBA, S.E., CARPENTER, S.B.**: Effect of shade on the growth, leaf morphology, and photosynthetic capacity of an American sycamore clone. - Castanea 45: 219-227, 1980.

*8452 - **DUBOIS, J.D., KAPUSTKA, L.A.**: Comparison of terrestrial and aquatic cyanobacteria to water stress effects on $N_2(C_2H_2)$ase activity. - Plant Physiol. 65(Suppl.): 8, 1980.

8453 - **DUBOIS, J.D., KAPUSTKA, L.A.**: Osmotic stress effects on the $N_2(C_2H_2)$ase activity on aquatic cyanobacteria. - Aquat. Bot. 11: 11-20, 1981.

*8454 - **DUDNIK, S.A., ANTONOV, A.V., PLESHKOV, K.K.**: Urozhaĭ ranneĭ kapusty v lesostepi USSR v zavisimosti ot rezhima orosheniya i gustoty posadki. [Early cabbage yield in forest-steppe of USSR in dependence on irrigation regime and stand density.] - Nauch.-tekh. Byull. (Khar'kov) 1980(11):11-20, 1980. [In R.]

8455 - **DUKE, N.C., BIRCH, W.R., WILLIAMS, W.T.**: Growth rings and rainfall correlations in a mangrove tree of the genus *Diospyros* (Ebenaceae). - Aust. J. Bot. 29: 135-142, 1981.

*8456 - **DULOV, S., DARZHANOV, K.**: S"otnoshenie mezhdu evapotranspiratsiyata na tsarevitsata i izparyaemostta. [Ratio between evapotranspiration of maize and evaporation.] - Rasteniev. Nauki. 17(1): 82-90, 1980. [In Bulg, ab: R, F.]

*8457 - **DÜRING, H., BROQUEDIS, M.**: Effects of abscisic acid and benzyladenine on irrigated and non-irrigated grapevines. - Scientia Hort. 13: 253-260, 1980.

8458 - **DUTEAU, J., GUILLOUX, M., GLORIES, Y., SEGUIN, G.**: Influence de l'alimentation en eau de la Vigne sur la teneur en sucres réducteurs, acides organiques et composés phénoliques des Raisins. - C.R. Acad. Sci. Paris Sér. III 292: 965-967, 1981.

8459 - **DUTZMANN, S., FORCHE, S., DÖLL, G., KRANZ, J.**: Licht- und rasterelektronenmikroskopischer Vergleich von Blattoberflächen verschiedener Sommergerstensorten. - Z. Pflanzenkrank. Pflanzenschutz 88: 518-524, 1981.

8460 - DWELLE, R.B., KLEINKOPF, G.E., PAVEK, J.J.: Stomatal conductance and gross photosynthesis of potato (*Solanum tuberosum* L.) as influenced by irradiance, temperature, and growth stage. - Potato Res. 24: 49-59, 1981.

8461 - DWELLE, R.B., KLEINKOPF, G.E., STEINHORST, R.K., PAVEK, J.J., HURLEY, P.J.: The influence of physiological processes on tuber yield of potato clones (*Solanum tuberosum* L.): Stomatal diffusive resistance, stomatal conductance, gross photosynthetic rate, leaf canopy, tissue nutrient levels, and tuber enzyme activities. - Potato Res. 24: 33-47, 1981.

8462 - DYCK, W.J., WEBBER, B.D., BARTON, P.G.: Soil-water nutrient concentrations after clearfelling and burning of *Pinus radiata*. - N. Zeal. J. Forest Sci. 11: 128-143, 1981.

8463 - DYLENOK, L.A., KHOTYLEVA, L.V., YATSEVICH, A.P.: Ispol'zovanie monosomnykh liniĭ yarovoĭ pshenitsy dlya izucheniya geneticheskogo kontrolya chastoty i razmerov ust'its flagovogo lista. [Use of monosomic spring wheat lines for studying the genetic control of stomatal frequency and sizes of the flag leaves.] - Dokl. Akad. Nauk Beloruss. SSR 25: 753-755, 1981. [In R, ab: E.]

*8464 - EBEL, F., KÄSTNER, A.: Notizen zur Morphologie und Blattanatomie einiger auf Serpentinböden vorkommender kubanischer Gehölze. - Wiss. Z. Univ. Halle 27 (3): 93-101, 1978.

8465 - ECK, H.V., MARTINEZ, T., WILSON, G.C.: Tall fescue and smooth bromegrass. I. Nitrogen and water requirements. - Agron. J. 73: 446-452, 1981.

8466 - ECK, H.V., WILSON, G.C., MARTINEZ, T.: Tall fescue and smooth bromegrass. II. Effects of nitrogen fertilization and irrigation regimes on quality. - Agron. J. 73: 453-456, 1981.

8467 - EDER, A., STICHLER, W., ZIEGLER, H.: Mechanismen der CO_2-Fixierung bei *Euphorbia trigona* Haw. und einigen *Pachypodium*-Arten. - Biochem. Physiol. Pflanz. 176: 1-12, 1981.

8468 - EDWARDS, W.R.N., JARVIS, P.G.: A new method of measuring water potential in tree stems by water injection. - Plant Cell Environ. 4: 463-465, 1981.

8469 - EGLEY, G.H., PAUL, R.N.,Jr.: Morphological observations on the early imbibition of water by *Sida spinosa* (Malvaceae) seed. - Amer. J. Bot. 68: 1056-1965, 1981.

*8470 - EHLERINGER, J., COOK, C.S.: Measurements of photosynthesis in the field: utility of the CO_2 depletion technique. - Plant Cell Environ. 3: 479-482, 1980.

8471 - EHLERINGER, J., MOONEY, H.A., GULMON, S.L., RUNDEL, P.W.: Parallel evolution of leaf pubescence in *Encelia* in coastal deserts of North and South America. - Oecologia 49: 38-41, 1981.

8472 - EHLERS, W., KHOSLA, B.K., KÖPKE, U., STÜLPNAGEL, R., BÖHM, W., BAEUMER, K.: Tillage effects on root development, water uptake and growth of oats. - Soil Tillage Res. 1: 19-34, 1980/1981.

*8473 - EHRET, D.L., JOLLIFFE, P.A.: A looped open system for plant gas exchange measurements. - Plant Physiol. 65 (Suppl.): 74, 1980.

*8474 - ELIÁŠ, P.: Vodný režím rastlín v opadavom listnatom lese. [Plant water relations in a deciduous broad-leaves forest.] - In: Zborník Referátov 3. Zjazdu Slovenskej Botanickej Spoločnosti Zvolen 1980. Pp. 223-225. Slovenská Botanická Spoločnost', Bratislava 1980. [In Slov, ab: R, E.]

8475 - ELIÁŠ, P.: Some ecophysiological leaf-characteristics of components of spring synuzium in temperate deciduous forests. - Biológia (Bratislava) 36: 841-849, 1981.

8476 - ELIÁŠ, P.: Visual estimation of leaf water stress in *Mercurialis perennis* L. - Biol. Plant. 23: 456-461, 1981.

8477 - ELLENBERG, H.: Ursachen des Vorkommens und Fehlens von Sukkulenten in den Trockengebieten der Erde. - Flora 171: 114-169, 1981.

8478 - ELLER, B.M., GLÄTTLI, R., FLACH, B.: Optische Eigenschaften und Pigmente von Sonnen- und Schattenblättern der Rotbuche (*Fagus silvatica* L.) und der Blut-buche (*Fagus silvatica* cv. Atropunicea). - Flora 171: 170-185, 1981.

8479 - ELLIS, M.A., FERREE, D.C.: How powdery mildew affects apple photosynthesis and transpiration. - Ohio Rep. 66 (5): 67-70, 1981.

8480 - ELLIS, M.A., FERREE, D.C., SPRING, D.E.: Photosynthesis, transpiration, and carbohydrate content of apple leaves infected by *Podosphaera leucotricha*. - Phytopathology 71: 392-395, 1981.

8481 - ELLIS, R.H., HONG, T.D., ROBERTS, E.H.: The influence of desiccation on cassava seed germination and longevity. - Ann. Bot. 47: 173-175, 1981.

8482 - ELMORE, C.D., McMICHAEL, B.L.: Proline accumulation by water and nitrogen stressed cotton. - Crop Sci. 21: 244-248, 1981.

*8483 - ELTHON, T.E., STEWART, C.R.: Proline oxidation in isolated corn mitochondria. - Plant Physiol. 65 (Suppl.): 8, 1980.

8484 - ENCIU, M., PLOAE, V., SIPOŞ, G.: Unele aspecte privind irigarea porumbului. [Some aspects of the irrigation of maize crops.] - An. Inst. Cercetări Cereale Plante tehnice Fundulea 47: 223-229, 1981. [In Roum, ab: R, E.]

*8485 - ENEVA, S.: Zavisimost mezhdu evapotranspiratsiyata na tsarevitsata i nyakoi meteorologichni faktori. [Correlation between maize evapotranspiration and some meteorological factors.] - Rasteniev. Nauki 17 (1): 98-102, 1980. [In Bulg, ab: E, R.]

*8486 - ENEVA, S.: Vliyanie na ogranichenoto vodosnabdyavane i mineralnoto torene v"rkhu dobiva na tsarevitsa za z"rno. [Effect of limited water supply and mineral fertilization on the yield of grain maize.] - Rasteniev. Nauki 17 (2): 74-81, 1980. [In Bulg, ab: R, E.]

*8487 - ENEVA, S.: V"rkhu nyakoi zavisimosti mezhdu vodata i dobiva pri napoyavane na tsarevitsa za z"rno na izluzhen chernozem-smolnitsa. [On some water-yield correlations of irrigated maize on leached chernozem-smolnitsa soils.] - Rasteniev. Nauki 17(4): 124-134, 1980. [In Bulg, ab: R, E.]

8488 - ENGLISH, M.J.: The uncertainty of crop models in irrigation optimization. - Trans. ASAE 24: 917-921,928, 1981.

*8489 - ENOCH, S., GLINKA, Z.: The effect of cells' turgidity on influx and efflux of K^+. - Plant Physiol. 65 (Suppl.): 61, 1980.

8490 - ENOCH, S., GLINKA, Z.: Changes in potassium fluxes in cells of carrot storage tissue related to turgor pressure. - Physiol. Plant. 53: 548-552, 1981.

*8491 - EPSTEIN, E., NORLYN, J.D., RUSH, D.W., KINGSBURY, R.W., KELLEY, D.B., CUNNINGHAM, G.A., WRONA, A.F.: Saline culture of crops: A genetic approach. - Science 210: 399-404, 1980.

*8492 - ESTEP, M.F., HOERING, T.C.: Biogeochemistry of the stable hydrogen isotopes.
        - Geochim. cosmochim. Acta 44: 1197-1206, 1980.

8493 - EVANS, D.D., SAMMIS, T.W., CABLE, D.R.: Actual evapotranspiration under de-
       sert conditions. - In: EVANS, D.D., THAMES, J.L. (ed.): Water in Desert
       Ecosystems. Pp. 195-218. Dowden, Hutchinson & Ross, Inc., Stroudsburg 1981.

8494 - EVANS, D.D., THAMES, J.L. (ed.): Water in Desert Ecosystems (US/IBP Synthesis
       Series 11). - Dowden, Hutchinson & Ross, Inc., Stroudsburg 1981.

8495 - EVANS, L.S., CURRY, T.M., LEWIN, K.F.: Responses of leaves of *Phaseolus
       vulgaris* L. to simulated acidic rain. - New Phytol. 88: 403-420, 1981.

8496 - EZE, J.M.O., DUMBROFF, E.B., THOMPSON, J.E.: Effects of moisture stress and
       senescence on the synthesis of abscisic acid in the primary leaves of bean.
       - Physiol. Plant. 51: 418-422, 1981.

8497 - FAGERIA, N.K., FILHO, M.P.B., GHEYI, H.R.: Avaliação de cultivares de arroz
       para tolerância à salinidade. [Screening rice cultivars for salinity tole-
       rance.] - Pesq. agropec. Brasil. 16: 677-681, 1981. [In Port, ab: E.]

*8498 - FAHEY, R.C., DI STEFANO, D.L., MEIER, G.P., BRYAN, R.N.: Role of hydration
        state and thiol-disulfide status in the control of thermal stability and
        protein synthesis in wheat embryo. - Plant Physiol. 65: 1062-1066, 1980.

*8499 - FALLOON, P.G., WHITE, J.G.H.: Development of reproductive structures in
        field peas (*Pisum sativum*) at different densities. - N. Zeal. J. agr. Res.
        23: 243-248, 1980.

8500 - FARAH, S.M.: An examination of the effects of water stress on leaf growth
       of crops of field beans (*Vicia faba* L.). 1. Crop growth and yield. - J. agr.
       Sci. 96: 327-336, 1981.

8501 - FECHNER, G.H., BURR, K.E., MYERS, J.F.: Effects of storage, temperature, and
       moisture stress on seed germination and early seedling development of trem-
       bling aspen. - Can. J. Forest Res. 11: 718-722, 1981.

8502 - FEDULOV, Yu.P., CHUVAEVA, A.D.: Vliyanie sveta na ustoĭchivost' list'ev
       ozimoĭ pshenitsy k deĭstviyu vysokoĭ temperatury i obezvozhivaniya.
       [Influence of light on resistance of winter wheat leaves to high temperature
       and dehydration.] - Fiziol. Rast. 28: 36-42, 1981. [In R, ab: E.]

8503 - FEINDT, F., MENDGEN, K., HEITEFUSS, R.: Der Einfluss der Spaltöffnungsweite
       und des Blattalters auf den Infektionserfolg von *Cercospora beticola* bei
       Zuckerrüben (*Beta vulgaris* L.) unterschiedlicher Anfälligkeit. - Phytopathol.
       Z. 101: 281-297, 1981.

*8504 - FERGUSON, A.R.: Xylem sap from *Actinidia chinensis*: Apparent differences in
        sap composition arising from the method of collection. - Ann. Bot. 46: 791
        -801, 1980.

8505 - FERGUSON, A.R., EISEMAN, J.A., DALE, J.R.: Xylem sap from *Actinidia chinensis*:
       Gradients in sap composition. - Ann. Bot. 48: 75-80, 1981.

*8506 - FERREE, D.C., HALL, F.R.: Effects of soil water stress and twospotted spider
        mites on net photosynthesis and transpiration of apple leaves. - Photosynthe-
        sis Res. 1: 189-197, 1980.

8507 - FERREE, D.C., HALL, F.R.: Influence of physical stress on photosynthesis and
       transpiration of apple leaves. - J. Amer. Soc. hort. Sci. 106: 348-351,
       1981.

8508 - FERRIER, J.M., DILLON, E.M., CHANG, A., DAINTY, J.: The effect of auxin on
cell volume elastic modulus in storage tissue of *Helianthus tuberosus*. -
Can. J. Bot. 59: 505-507, 1981.

8509 - FETCHER, N.: Leaf size and leaf temperature in tropical vines. - Amer. Natur.
117: 1011-1014, 1981.

*8510 - FILIPPOV, L.A.: Pribor dlya opredeleniya transpiratsii list'ev pri pomoshchi
vodopogloshchayushchikh reagentov. [A device for measuring transpiration
with the aid of water-absorbing substances.] - Fiziol. Rast. 27: 1304-1307,
1980. [In R, ab: E.]

8511 - FISCHER, R.A.: Optimizing the use of water and nitrogen through breeding of
crops. - Plant Soil 58: 249-278, 1981. Also in: MONTEITH, J., WEBB, C. (ed.):
Soil Water and Nitrogen in Mediterranean-Type Environments. Development in
Plant and Soil Sciences. Volume 1. Pp. 249-278. Martinus Nijhoff / Dr. W. Junk
Publishers, The Hague - Boston - London 1981.

8512 - FISCHER, R.A., BIDINGER, F., SYME, J.R., WALL, P.C.: Leaf photosynthesis,
leaf permeability, crop growth, and yield of short spring wheat genotypes
under irrigation. - Crop Sci. 21: 367-373, 1981.

8513 - FISCHER, W., LISTE, H.-J., BELEITES, F.: Untersuchungen zum Einfluss speziali-
sierter Fruchtfolgen und der Beregnung auf die Ertragsentwicklung bei Zuckerrüben. - Arch. Acker- Pflanzenbau Bodenk. 25: 109-116, 1981.

8514 - FISCUS, E.L.: Effects of abscisic acid on the hydraulic conductance of and
the total ion transport through *Phaseolus* root systems. - Plant Physiol. 68:
169-174, 1981.

8515 - FISHER, M.J., CHARLES-EDWARDS, D.A., LUDLOW, M.M.: An analysis of the effects
of repeated short-term soil water deficits on stomatal conductance to carbon
dioxide and leaf photosynthesis by the legume *Macroptilium atropurpureum* cv.
Siratro. - Aust. J. Plant Physiol. 8: 347-357, 1981.

8516 - FLORE, J.A., BUKOVAC, M.J.: Pesticide effects on the plant cuticle: IV. The
effect of EPTC on the permeability of cabbage, bean, and sugar beet cuticle.
- J. Amer. Soc. hort. Sci. 106: 189-193, 1981.

8517 - FLOROV, R.J., STOJANOV, Zh.V.: Pflanzenatmung und Transpiration bei verschie-
denen Wassergehalt unter thermodynamischen Aspekten. - In: UNGER, K., STÖCKER,
G. (ed.): Biophysikalische Ökologie und Ökosystemforschung. Pp. 69-80.
Akademie-Verlag, Berlin 1981.

8518 - FLOWERS, T.J., YEO, A.R.: Variability in the resistance of sodium chloride
salinity within rice (*Oryza sativa* L.) varieties. - New Phytol. 88: 363-373,
1981.

8519 - FOK, M.V., BORISOV, A.Yu.: Rol' vody v stabilizatsii razdelennykh zaryadov
v pervichnom aket fotosinteza. [Model of water participation in stabilization
of charges in the primary photosynthesis events.] - Mol. Biol. 15: 575-582,
1981. [In R, ab: E.]

8520 - FOKEEV, P.M., MURAVLEV, A.P., KOLCHINA, N.A.: Biologicheskie problemy oro-
shaemogo zemledeliya (na primere Povolzh'ya). [Biological problems of irrigated farming (on example of Povolzh'e).]- Sel'skokhoz. Biol. 16: 500-509,
1981. [In R, ab: E.]

8521 - FORD, C.W.: A new lactone from water-stressed chickpea. - Phytochemistry 20:
2019-2020, 1981.

8522 - FORD, C.W., WILSON, J.R.: Changes in levels of solutes during osmotic adjustment to water stress in leaves of four tropical pasture species. - Aust. J.
Plant Physiol. 8: 77-91, 1981.

8523 - FORD, E.D.: Can we model xylem production by conifers? - Studia forest. Suecica 160 (LINDER, S. (ed.): Understanding and Predicting Tree Growth.): 19-
29, 1981.

*8524 - FORD, E.D., MILNE, R.: Assessing plant response to the weather. - In: GRACE,
J., FORD, E.D., JARVIS, P.G. (ed.): Plants and their Atmospheric Environment.
Pp. 333-362. Blackwell Scientific Publications, Oxford - London - Edinburgh
- Boston - Melbourne 1980.

*8525 - FORSETH, I., EHLERINGER, J.R.: Solar tracking response to drought in a desert
annual. - Oecologia 44: 159-163, 1980.

*8526 - FÖRSTER, H., BUCHENAUER, H., GROSSMANN, F.: Nebenwirkungen der systemischen
Fungizide Triadimefon und Triadimenol auf Gerstenpflanzen. III. Weitere
Beeinflussungen des Stoffwechsels. - Z. Pflanzenkrank. Pflanzenschutz 87:
717-730, 1980.

8527 - FOWKES, N.D., LANDSBERG, J.J.: Optimal root systems in terms of water uptake
and movement. - In: ROSE, D.A., CHARLES-EDWARDS, D.A. (ed.): Mathematics and
Plant Physiology. Pp. 109-125. Academic Press, London - New York - Toronto -
Sydney - San Francisco 1981.

*8528 - FOWLER, D.: Turbulent transfer of sulphur dioxide to cereals: a case study.
- In: GRACE, J., FORD, E.D., JARVIS, P.G. (ed.): Plants and their Atmospheric
Environment. Pp. 139-146. Blackwell Scientific Publications, Oxford - London
- Edinburgh - Boston - Melbourne 1980.

8529 - FOWLER, D.B.: Fall growth and cold acclimation of winter wheat and rye on
saline soils. - Can. J. Plant Sci. 61: 225-230, 1981.

8530 - FOX, J.F., HARRISON, A.T.: Habitat assortment of sexes and water balance in
a dioecious grass. - Oecologia 49: 233-235, 1981.

8531 - FRANK, A.B.: Effect of leaf age and position on photosynthesis and stomatal
conductance of forage grasses. - Agron. J. 73: 70-74, 1981.

*8532 - FRANZ, H. (ed.): Untersuchungen an Alpinen Böden in den Hohen Tauern 1974-
1978 Stoffdynamik und Wasserhaushalt. (Veröffentlichungen des Österreichis-
chen MaB-Hochgebirgsprogramms Hohe Tauern, Band 3). - Universitätsverlag
Wagner, Innsbruck 1980.

8533 - FREER-SMITH, P.H., WILLMER, C.M.: Guard cell metabolism in epidermis of
Commelina communis L. during stomatal opening and closing. - J. exp. Bot.
32: 535-543, 1981.

8534 - FRIC, V., MAKOVEC, K., SOUHRADOVÁ, A., LOUKOTA, M.: Dynamika vadnutí nadzem-
ních orgánů chmelové rostliny po skliźňové dekapitaci. [The dynamics of the
wilting of the above-ground organs of hop plants after harvest decapitation.]
- Rost. Výroba (Praha) 27: 97-102, 1981. [In Czech, ab: R, E, G.]

8535 - FRIEDMAN, J., STEIN, Z., RUSHKIN, E.: Drought tolerance of germinating seeds
and young seedlings of *Anastatica hierochuntica* L. - Oecologia 51: 400-403,
1981.

*8536 - FRIEDRICH, J.W., HUFFAKER, R.C.: Photosynthesis, leaf resistances, and
ribulose-1,5-bisphosphate carboxylase degradation in senescing barley leaves.
- Plant Physiol. 65: 1103-1107, 1980.

*8537 - FRIEDRICH, J.W., HUFFAKER, R.C.: Photosynthesis, ribulose-1,5-bisphosphate carboxylase, and leaf resistances in barley. - Plant Physiol. 65 (Suppl.): 72, 1980.

*8538 - FRITSCHEN, L.J., WALKER, R.B., HSIA, J.: Energy balance of an isolated Scots pine. - Int. J. Biometeorol. 24: 293-300, 1980.

8539 - FUCHIGAMI, L.H., CHENG, T.Y., SOELDNER, A.: Abaxial transpiration and water loss in aseptically cultured plum. - J. Amer. Soc. hort. Sci. 106: 519-522, 1981.

8540 - FUHRER, J., ERISMANN, K.H.: On the use of a field exposure system to estimate air pollution levels. - Int. J. environ. Studies 16: 85-89, 1981.

*8541 - FUHRMANN, U.: Beziehungen zwischen Witterung und Jahresertrag auf Niederungsgrünland - ein Beitrag zur Erarbeitung von Ertragsrichtwerten. - Arch. Acker-Pflanzenbau Bodenk. 22: 593-599, 1978.

8542 - FUHRMANN, U., HAGER, U.: Verbesserte Ertragsberechnung für Niederungsgrasland durch quantifizierte meteorologische Daten. - Arch. Acker- Pflanzenbau Bodenk. 25: 573-578, 1981.

8543 - FUKUTOKU, Y., YAMADA, Y.: Diurnal changes in water potential and free amino acid contents of water-stressed and non-stressed soybean plants. - Soil Sci. Plant Nutr. 27: 195-204, 1981.

8544 - FUKUTOKU, Y., YAMADA, Y.: Sources of proline-nitrogen in water-stressed soybean (Glycine max L.) I. Protein metabolism and proline accumulation. - Plant Cell Physiol. 22: 1397-1404, 1981.

*8545 - FURUKAWA, A., ISODA, O., IWAKI, H., TOTSUKA, T.: Interspecific difference in resistance to sulfur dioxide. - Res. Rep. nat. Inst. environ. Studies 11 (Studies on the Effects of Air Pollutants on Plants Mechanisms of Phytotoxicity): 113-126, 1980.

*8546 - FURUKAWA, A., NATORI, T., TOTSUKA, T.: The effect of SO_2 on net photosynthesis in sunflower leaf. - Res. Rep. nat. Inst. environ. Studies 11 (Studies on the Effects of Air Pollutants on Plants and Mechanisms of Phytotoxicity): 1-8, 1980.

8547 - GACHECHILADZE, N.D., KORZINNIKOV, Yu.S., GLAZUNOVA, E.M., YUSUFBEKOV, Kh.Yu., BONDAR', V.V., KRYMSKAYA, N.B., POTAPOVA, I.M.: Biokhimicheskaya i morfologicheskaya kharakteristika form oblepikhi krushinovidnoĭ, proirastayushcheĭ na Zapadnom Pamire. [Biochemical and morphological characteristics of forms of Hippophaë rhamnoides growing in the Western Pamir.] - Rast. Resursy 17 (1): 37-42, 1981. [In R.]

*8548 - GALES, K.: Effects of water supply on partitioning of dry matter between roots and shoots in Lolium perenne. - J. appl. Ecol. 16: 863-877, 1979.

8549 - GALES, K., WILSON, N.J.: Effects of water shortage on the yield of winter wheat. - Ann. appl. Biol. 99: 323-334, 1981.

*8550 - GALLI, M.G., MIRACCA, P., SPARVOLI, E.: Lack of inhibiting effects of abscisic acid on seeds of Haplopappus gracilis (Nutt.) Gray pregerminated in water for short times. - J. exp. Bot. 31: 763-770, 1980.

*8551 - GAMAYUN, I.M.: Razvitie, rost i produktivnost' srednespelogo tomata v zavisimosti ot rezhimov orosheniya. [Development, growth, and productivity of middle-ripening tomato considering the irrigation regimes.] - Sel'skokhoz. Biol. 15: 141-142, 1980. [In R.]

8552 - GANSKOPP, D.C., BEDELL, T.E.: An assessment of vigor and production of range
       grasses following drought. - J. Range Manage. 34: 137-141, 1981.

8553 - GARDNER, B.R., BLAD, B.L., GARRITY, D.P., WATTS, D.G.: Relationships between
       crop temperature, grain yield, evapotranspiration and phenological develop-
       ment in two hybrids of moisture stressed sorghum. - Irrig. Sci. 2: 213-224,
       1981.

8554 - GARDNER, B.R., BLAD, B.L., MAURER, R.E., WATTS, D.G.: Relationship between
       crop temperature and the physiological and phenological development of
       differentially irrigated corn. - Agron. J. 73: 743-747, 1981.

8555 - GARDNER, B.R., BLAD, B.L., WATTS, D.G.: Plant and air temperatures in diffe-
       rentially-irrigated corn. - Agr. Meteorol. 25: 207-217, 1981.

8556 - GARG, B.K., KATHJU, S., LAHIRI, A.N., VYAS, S.P.: Drought resistance in
       pearl millet. - Biol. Plant. 23: 182-185, 1981.

*8557 - GARG, K.K., VARSHNEY, C.K.: Effect of air pollution on the leaf epidermis at
       the submicroscopic level. - Experientia 36: 1364-1366, 1980.

*8558 - GASH, J.H.C.: An analytical model of rainfall interception by forests. -
       Quart. J. roy. meteorol. Soc. 105: 43-55, 1979.

*8559 - GASH, J.H.C., WRIGHT, I.R., LLOYD, C.R.: Comparative estimates of intercep-
       tion loss from three coniferous forests in Great Britain. - J. Hydrol. 48:
       89-105, 1980.

*8560 - GATES, C.T., SANDLAND, R.L.: Appraisal by numerical and statistical techni-
       ques of the interaction of moisture stress and phosphorus in the development
       of the tropical legume *Macroptilium atropurpureum* (DC.) Urb. - Aust. J. Bot.
       28: 621-631, 1980.

8561 - GAUSMAN, H.W., MENGES, R.M., RICHARDSON, A.J., WALTER, H., RODRIGUEZ, R.R.,
       TAMEZ, S.: Optical parameters of leaves of seven weed species. - Weed Sci.
       29: 24-26, 1981.

8562 - GAY, L.W.: Potential evapotranspiration for deserts. - In: EVANS, D.D.,
       THAMES, J.L. (ed.): Water in Desert Ecosystems. Pp. 172-194. Dowden, Hutchin-
       son & Ross, Inc., Stroudsburg 1981.

*8563 - GAY, L.W., FRITSCHEN, L.J.: An energy budget analysis of water use by salt-
       cedar. - Water Resour. Res. 15: 1589-1592, 1979.

*8564 - GEHRING, J.M., LEWIS, A.J.,III: Effect of hydrogel on wilting and moisture
       stress of bedding plants. - J. Amer. Soc. hort. Sci. 105: 511-513, 1980.

8565 - GEISLER, G., DAROUSSIS, N.: Einfluss von Porengrössenverteilung, Wasserange-
       bot und luftführendem Porenvolumen auf morphologische Eigenschaften des
       Wurzelsystems sowie das Wachstum von Mais, Ackerbohne und Gerste. - Z. Acker-
       Pflanzenbau 150: 457-473, 1981.

8566 - GEL'TMAN, V.S., PYATROŬ, Ya.G., RÉUTSKI, V.G.: Zakanamernastsi vodnaga zhyŭ-
       lennya fitatsènozaŭ. [Regularities of water supply to the phytocenoses.] -
       Vostsi Akad. Navuk Bolarus. SSR, Scr. biyal. Navuk 1981 (2): 44-49, 156,
       1981. [In Belorus, ab: E, R.]

8567 - GENKEL', P.A., TIKHOMIROVA, E.V.: Vliyanie zakalivaniya protiv zasukhi na
       soderzhanie belka i aktivnost' glutaminsintetazy u kul'turnykh zlakov.
       [Effect of drought-hardening on protein content and glutamine synthetase
       activity in cereals.] - Fiziol. Rast. 28: 334-339, 1981. [In R, ab: E.]

8568 - **GERALDSON, C.M.:** Relevance of water and fertilizer to production efficiency of tomatoes and pepper. - Proc. Florida State hort. Soc. 92: 74-76, 1979.

*8569 - **GERTSYK, V.V., RODE, A.A.:** O sootnoshenii mezhdu kolichestvom atmosfernykh osadkov i velichinoĭ raskhoda vlagi celinnymi moshchnymi chernozemami pod nekosimoĭ step'yu i dubovym lesom. [Correlation between the amount of precipitation and the quantity of moisture discharge from deep chernozems under uncut steppe and oak forest.] - Pochvovedenie 1979 (11): 140-147, 1979. [In R, ab: E.]

*8570 - **GERTSYK, V.V., RODE, A.A.:** Statisticheskaya kharakteristika ĕlementov balansa vlagi chernozemov pod nekosimoĭ step'yu i dubovym lesom. [Statistical characteristic of moisture balance elements in virgin chernozems under uncut steppe and oak forest.] - Pochvovedenie 1980 (7): 101-111, 1980. [In R, ab: E.]

8571 - **GIBBON, D.:** Rainfed farming systems in the Mediterranean region. - Plant Soil 58: 59-80, 1981. Also in: MONTEITH, J., WEBB, C. (ed.): Soil Water and Nitrogen in Mediterranean-Type Environments. Development in Plant and Soil Sciences. Volume 1. Pp. 59-80. Martinus Nijhoff / Dr. W. Junk Publishers, The Hague - Boston - London 1981.

*8572 - **GILREATH, P.R., BUCHANAN, D.W.:** Evaporative cooling with overhead sprinkling for rest termination of peach trees. - Proc. Florida State hort. Soc. 92: 262-264, 1979.

8573 - **GINZBURG, M.:** Measurements of ion concentrations and fluxes in *Dunaliella parva*. - J. exp. Bot. 32: 321-332, 1981.

8574 - **GINZBURG, M.:** Measurements of ion concentrations in *Dunaliella parva* subjected to hypertonic shock. - J. exp. Bot. 32: 333-340, 1981.

*8575 - **GISI, U., ZENTMYER, G.A.:** Mechanism of zoospore release in *Phytophthora* and *Pythium*. - Exp. Mycol. 4: 362-377, 1980.

8576 - **GIURGEVICH, J.R., DUNN, E.L.:** A comparative analysis of the CO_2 and water vapor responses of two *Spartina* species from Georgia coastal marshes. - Estuarine coastal Shelf Sci. 12: 561-568, 1981.

*8577 - **GLINKA, Z.:** Abscisic acid promotes both volume flow and ion release to the xylem in sunflower roots. - Plant Physiol. 65: 537-540, 1980.

8578 - **GLOVER, L.J., ALLEN, A.L., PORTER, O.A.:** Irrigation studies with soybeans in southeast Arkansas. - Arkansas Farm Res. 30 (4): 7, 1981.

8579 - **GOBLE-GARRATT, E.M., BELL, D.T., LONERAGAN, W.A.:** Floristic and leaf structure patterns along a shallow elevation gradient. - Aust. J. Bot. 29: 329-347, 1981.

*8580 - **GOLDMAN, A., DOVRAT, A.:** Irrigation regime and honeybee activity as related to seed yield in alfalfa. - Agron. J. 72: 961-965, 1980.

8581 - **GOLTZ, S.M., BENOIT, G., SCHIMMELPFENNIG, H.:** New circuitry for measuring soil water matric potential with moisture blocks. - Agr. Meteorol. 24: 75-82, 1981.

*8582 - **GOMEZ, J.B., HAMZAH, S.B.:** Variations in leaf morphology and anatomy between clones of *Hevea*. - J. Rubber Res. Inst. Malaysia 28: 157-172, 1980.

8583 - **GONCHAROVA, Ê.A., UDOVENKO, G.V.:** Rostovaya i gormonal'naya aktivnost' list'ev i plodov v svyazi s ikh vzaimodeĭstviem pri raznoĭ vodoobespechennosti rasteniĭ. [The growth and hormonal activity of leaves and fruits and their relationships under different water supply of plants.] - Fiziol. Biokhim. kul't. Rast. 13: 125-131, 1981. [In R, ab: E.]

8584 - GOODMAN, R.N., WHITE, J.A.: Xylem parenchyma plasmolysis and vessel wall disorientation caused by *Erwinia amylovora*. - Phytopathology 71: 844-852, 1981.

*8585 - GOPALAKRISHNA PILLAI, K., DE, R.: Growth and grain yield of rice variety *Jaya* at different levels and timings of nitrogen application under two systems of water management. - Proc. Ind. Acad. Sci., Plant Sci. 89: 243-256, 1980.

*8586 - GOPALAKRISHNA PILLAI, K., DE, R.: Nutrient uptake of rice variety *Jaya* at different levels and timings of nitrogen application under two systems of water management. - Proc. Ind. Acad. Sci., Plant Sci. 89: 257-267, 1980.

*8587 - GORDON, D.M., BIRCH, P.B., MCCOMB, A.J.: The effect of light, temperature and salinity on photosynthetic rates of an estuarine *Cladophora*. - Bot. Mar. 23: 749-755, 1980.

8588 - GORHAM, J., HUGHES, I., WYN JONES, R.G.: Low-molecular-weight carbohydrates in some salt-stressed plants. - Physiol. Plant. 53: 27-33, 1981.

8589 - GOVINDJEE, DOWNTON, W.J.S., FORK, D.C., ARMOND, P.A.: Chlorophyll a fluorescence transient as an indicator of water potential of leaves. - Plant Sci. Lett. 20: 191-194, 1981.

*8590 - GRACE, J.: Some effects of wind on plants. - In: GRACE, J., FORD, E.D., JARVIS, P.G. (ed.): Plants and their Atmospheric Environment. Pp. 31-56. Blackwell Scientific Publications, Oxford - London - Edinburgh - Boston - Melbourne 1980.

*8591 - GRACE, J., FORD, E.D., JARVIS, P.G. (ed.): Plants and their Atmospheric Environment (The 21st Symposium of British Ecological Society, Edinburgh 26-30 March 1979). - Blackwell Scientific Publications, Oxford - London - Edinburgh - Boston - Melbourne 1980.

8592 - GRANDIN, M.: Action du NaCl et de la saturation en eau du sol sur le développement et les teneurs en glucides, minéraux et Na de *Glaux maritima* L. - Acta Oecol. - Oecol Plant. 2: 23-29, 1981.

8593 - GRANIER, A., LÉVY, G.: Influence des conditions d'engorgement du sol sur l'évolution de l'état hydrique de jeunes plants d'Epicéa (*Picea abies* L). - Ann. Sci. forest. 38: 179-198, 1981.

*8594 - GRAVES, C.R., McCUTCHEN, T., JEFFERY, L.S., OVERTON, J.R., HAYES, R.M.: Soybean-wheat cropping systems: Evaluation of planting methods, varieties, row spacings, and weed control. - Univ. Tennessee Agr. Exp. Sta. Bull. 597: 1-42, 1980.

*8595 - GREEN, J.M., WILLIAMS, G.J.: The effect of temperature on the gas exchange of temperate region cacti. - Plant Physiol. 65 (Suppl.): 48, 1980.

8596 - GREEN, T.G.A., SNELGAR, W.P.: Carbon dioxide exchange in lichens. Relationship between net photosynthetic rate and CO_2 concentration. - Plant Physiol. 68: 199-201, 1981.

8597 - GREEN, T.G.A., SNELGAR, W.P.: Carbon dioxide exchange in lichens: Partition of total CO_2 resistances at different thallus water contents into transport and carboxylation components. - Physiol. Plant. 52: 411-416, 1981.

8598 - GREEN, T.G.A., SNELGAR, W.P., BROWN, D.H.: Carbon dioxide exchange in lichens. Carbon dioxide exchange through the cyphellate lower cortex of *Sticta latifrons* Rich. - New Phytol. 88: 421-426, 1981.

8599 - GREENWOOD, D.J.: Crop response to agronomic practice. - In: ROSE, D.A., CHARLES-EDWARDS, D.A. (ed.): Mathematics and Plant Physiology. Pp. 195-216. Academic press, London - New York - Toronto - Sydney - San Francisco 1981.

8600 - GREENWOOD, M.S.: Reproductive development in loblolly pine. II. The effect of age, gibberellin plus water stress and out-of-phase dormancy on long shoot growth behavior. - Amer. J. Bot. 68: 1184-1190, 1981.

*8601 - GREGERSEN, A.K.: Vand og kvaelstofgødning til flerårigt graes og kløvergraes. [Water and nitrogen supply for pure grass and clovergrass.] - Tidsskr. Planteavl 84: 191-208, 1980. [In Dan, ab: E.]

8602 - GRIEVE, P.W., POVEY, M.J.W.: Evidence for osmotic dehydration. Theory of freeze damage. - J. Sci. Food Agr. 32: 96-98, 1981.

8603 - GRIFFITHS, J.H., JARVIS, P.G.: A null balance carbon dioxide and water vapour porometer. - J. exp. Bot. 32: 1157-1168, 1981.

*8604 - GRIME, J.P.: Plant Strategies and Vegetation Processes. - John Wiley & Sons, Chichester - New York - Brisbane - Toronto 1979.

8605 - GRINCHENKO, A.L., NAZARENKO, O.A.: Vliyanie retardantov na rost i produktiv-nost' kukuruzy i sorgo. [Effect of retardants on maize and sorghum growth and productivity.] - Fiziol. Biokhim. kul't. Rast. 13: 451-457, 1981. [In R, ab: E.]

8606 - GROSS, L.J.: On the dynamics of internal leaf carbon dioxide uptake. - J. Math. Biol. 11: 181-191, 1981.

8607 - GROYA, F.L., SHEAFFER, C.C.: Establishment of sod-seeded alfalfa at various levels of soil moisture and grass competition. - Agron. J. 73: 560-565, 1981.

8608 - GRZESIAK, S., ROOD, S.B., FREYMAN, S., MAJOR, D.J.: Growth of corn seedlings: Effects of night temperature under optimum soil moisture or under drought conditions. - Can. J. Plant Sci. 61: 871-877, 1981.

8609 - GUBBELS, G.H., KENASCHUK, E.O.: Desiccation of flax with diquat. - Can. J. Plant Sci. 61: 575-581, 1981.

*8610 - GUEDES, A.C., CANTLIFFE, D.J.: Germination of lettuce seeds at high tempe-rature after seed priming. - J. Amer. Soc. hort. Sci. 105: 777-781, 1980.

8611 - GUENNELON, R., CABIBEL, B.: Influence de l'activité du système racinaire de pommiers sur la répartition des solutés en irrigation localisée. - Agronomie 1: 323-330, 1981.

*8612 - GUGGINO, S., GUTKNECHT, J.: Turgor regulation in *Valonia macrophysa* after acute hyposmotic shock. - In: SPANSWICK, R.M., LUCAS, W.J., DAINTY, J. (ed.): Plant Membrane Transport: Current Conceptual Issues. Pp. 495-496. Elsevier / North-Holland Biomedical Press, Amsterdam - New York - Oxford 1980.

8613 - GUINN, G., MAUNEY, J.R., FRY, K.E.: Irrigation scheduling and plant popula-tion effects on growth, bloom rates, boll abscission, and yield of cotton. - Agron. J. 73: 529-534, 1981.

8614 - GULYAEV, B.I.: Voprosy kolichestvennogo opisaniya rostovykh funktsiĭ rasteniĭ. [Problems of a quatitative description of plant growth functions.] - Fiziol. Biokhim. kul't. Rast. 13: 227-238, 1981. [In R, ab: E.]

*8615 - GULYAEV, B.I., SLUKHAĬ, S.I., LIKHOLAT, D.A., PETRENKO, N.I.: Vliyanie urovnya kaliĭnogo pitaniya na fotosintez, dykhanie i diffuzionnoe sopro-tivlenie list'ev kukuruzy. [Effect of potassium nutrition level on photo-synthesis, respiration and diffusion resistance of maize leaves.] - Dokl. Akad. Nauk Ukr. SSR, Ser. B 1978(11): 1045-1048, 1978. [In R, ab: E.]

8616 - GUMBS, F.A., SIMPSON, L.A.: Influence of flooding and soil moisture content on elongation of sugar cane in Trinidat. - Exp. Agr. 17: 403-406, 1981.

*8617 - GUPTA, J.P.: Effect of mulches on moisture and thermal regimes of soil and yield of pearl millet. - Ann. Arid Zone 19: 132-138, 1980.

*8618 - GUPTA, J.P., AGGARWAL, R.K.: Effect of asphalt sub-surface moisture on water characteristics and productivity of sand soil. - Ann. Arid Zone 19: 445-450, 1980.

8619 - GUPTA, R.K.: Effects of various rates of desiccation on photosynthesis and photosynthates leakage in young and aged tissues of the moss *Fissidens adianthoides* (Hedw.). - Photosynthetica 15: 347-350, 1981.

*8620 - GUPTA, V., LAMBA, L.C.: Structure and development of pericarp stomata in *Papaveraceae*. - J. Sci. Res. 2: 19-22, 1980.

*8621 - GUTKNECHT, J., HASTINGS, D.F., BISSON, M.A.: Ion transport and turgor pressure regulation in giant algal cells. - In: GIEBISCH, G., TOSTESON, D.C., USSING, H.H. (ed.): Membrane Transport in Biology. Pp. 125-174. Springer -Verlag, Berlin - Heidelberg - New York 1978.

*8622 - GUY, C.L., YELENOSKY, G., SWEET, H.C.: Light exposure and soluble sugars in citrus frost hardiness. - Florida Sci. 43: 256-268, 1980.

*8623 - HÄCKEL, H.: Neues über die elektrische Methode zur Messung der Benetzungsdauer unmittelbar an der Pflanze. - Agr. Meteorol. 22: 113-119, 1980.

8624 - HAEDER, H.E., BERINGER, H.: Influence of potassium nutrition and water stress on the content of abscisic acid in grains and flag leaves of wheat during grain development. - J. Sci. Food Agr. 32: 552-556, 1981.

8625 - HALL, A.E.: Adaptation of annual plants to drought in relation to improvements in cultivars. - HortScience 16: 37-38, 1981.

8626 - HALL, A.E., GRANTZ, D.A.: Drought resistance of cowpea improved by selecting for early apperance of mature pods. - Crop Sci. 21: 461-464, 1981.

8627 - HALL, A.J., LEMCOFF, J.H., TRAPANI, N.: Water stress before and during flowering in maize and its effects on yield, its components, and their determinants. - Maydica 26: 19-38, 1981.

8628 - HALL, H.K., McWHA, J.A.: Effects of abscisic acid on growth of wheat (*Triticum aestivum* L.). - Ann. Bot. 47: 427-433, 1981.

*8629 - HALL, S.L., McPHERSON, J.K.: Geographic distribution of two species of oaks in Oklahoma in relation to seasonal water potential and transpiration rates. - Southwestern Natur. 25: 283-295, 1980.

*8630 - HALLDIN, S., GRIP, H., JANSSON, P.-E., LINDGREN, Å.: Micrometeorology and hydrology of pine forest ecosystems. II. Theory and models. - Ecol. Bull. (Stockholm) 32 (PERSSON, T. (ed.): Structure and Function of Northern Coniferous Forests - An Ecosystem Study.): 463-503, 1980.

8631 - HALTERLEIN, A.J., SCIUMBATO, G.L., BARRENTINE, W.L.: Use of plant desiccants to control cucumber fruit rot. - HortScience 16: 189-190, 1981.

8632 - HALVA, E., LESÁK, J., PAVLÍČEK, A.: Produktivnost lučních ekosystémů ovlivněných vodohospodářskými úpravami jižní Moravy. [Productivity of meadow ecosystems affected by water-management practices in south Moravia.] - Rost. Výroba (Praha) 27: 1139-1145, 1981. [In Czech, ab: R, E, G.]

*8633 - HAMABATA, A., MARTÍNEZ-C, J.L.: Some responses to hydration level during
        germination of *Triticum aestivum* V. Potam. - Plant Physiol. 65 (Suppl.): 102,
        1980.

 8634 - HAMER, P.J.C.: The effects of evaporative cooling on apple bud development
        and frost resistance. - J. hort. Sci. 56: 107-112, 1981.

 8635 - HANCOCK, J.G.: Osmotic conditions influence estimation of passive permeability
        changes in diseased tissues. - Physiol. Plant Pathol. 18: 117-122, 1981.

*8636 - HANSEN, D.H.: Physiology and microclimate in a hemiparasite *Castilleja chro-
        mosa* (*Scrophulariaceae*). - Amer. J. Bot. 66: 477-484, 1979.

 8637 - HANSEN, V.: Evapotranspirasjon, - fordamping fra vekster. [Evapotranspiration,
        - evaporation from vegetative surfaces.] - Meld. Norg. Landbrukshøgsk. 60
        (2): 1-12, 1981. [In Norg, ab: E.]

 8638 - HANSEN, V.: Advection and evapotranspiration. - Meld. Norg. Landbrukshøgsk.
        60(21): 2-7, 1981.

*8639 - HANSON, A.D., SCOTT, N.A.: Betaine synthesis from [^{14}C] precursors in water
        -stressed barley leaves. - Plant Physiol. 65 (Suppl.): 8, 1980.

*8640 - HARBAUGH, B.K., WILFRET, G.J.: Spray chrysanthemum production with controlled
        -release fertilizer and trickle irrigation. - J. Amer. Soc. hort. Sci. 105:
        367-371, 1980.

 8641 - HARDIE, K., LEYTON, L.: The influence of vesicular-arbuscular mycorrhiza on
        growth and water relations of red clover. I. In phosphate deficient soil. -
        New Phytol. 89: 599-608, 1981.

*8642 - HARI, P.: The dynamics of metabolism in a plant community. - Flora 170:
        28-50, 1980.

*8643 - HARMS, W.R., SCHREUDER, H.T., HOOK, D.D., BROWN, C.L., SHROPSHIRE, F.W.:
        The effects of flooding on the swamp forest in lake Ocklawaha, Florida. -
        Ecology 61: 1412-1421, 1980.

 8644 - HARRIS, G.A., CAMPBELL, G.S.: Morphological and physiological characteristics
        of desert plants. - In: EVANS, D.D., THAMES, J.L. (ed.): Water in Desert
        Ecosystems. Pp. 59-74. Dowden, Hutchinson & Ross, Inc., Stroudsburg 1981.

 8645 - HARRIS, M.J., HEATH, R.L.: Ozone sensitivity in sweet corn (*Zea mays* L.)
        plants: A possible relationship to water balance. - Plant Physiol. 68:
        885-890, 1981.

 8646 - HASPEL-HORVATOVIČ, E., HOLÚBKOVÁ, B.: Experimental studies of chlorophyll
        - water relations. - Phytopathol. Z. 100: 340-346, 1981.

*8647 - HAWKER, J.S.: Invertases from leaves of *Phaseolus vulgaris* plants grown on
        nutrient solutions containing NaCl. - Aust. J. Plant Physiol. 7: 67-72,
        1980.

*8648 - HAWKER, J.S., BUTTROSE, M.S.: Development of the almond nut (*Prunus dulcis*
        (Mill.) D.A. Webb). Anatomy and chemical composition of fruit parts from
        anthesis to maturity. - Ann. Bot. 46: 313-321, 1980.

 8649 - HAWKINS, C.D.B., LISTER, G.R., FINK, R.P., VIDAVER, W.E.: Short-term pigment
        changes in Norway spruce needles. - Physiol. Plant. 51: 175-180, 1981.

 8650 - HAYGOOD, R.A., STRIDER, D.L.: Influence of moisture and inoculum concentra-
        tion on infection of *Philodendron selloun* by *Erwinia chrysanthemi*. - Plant
        Dis. 65: 727-728, 1981.

8651 - HAYHOE, H.: Analysis of a diffusion model for plant root growth and an application to plant soil-water uptake. - Soil Sci. 131: 334-343, 1981.

8652 - HAYNES, R., HERRING, S.: Performance of drip-irrigated bell peppers. - Arkansas Farm Res. 30: 12, 1981.

8653 - HEARN, A.B., CONSTABLE, G.A.: Irrigation for crops in a sub-humid environment. V. Stress day analysis for soybeans and an economic evaluation of
strategies. - Irrig. Sci. 3: 1-15, 1981.

8654 - HEATH, O.V.S., MEIDNER, H.: Feedback processes in the opening of leaf stomata
in light. - Proc. roy. Soc. London, Ser. B 213: 161-170, 1981.

*8655 - HEERKLOSS, B., BARTOLOMAEUS, W.: Experimentelle Arbeiten zum Keimungsverhal-
ten von Kulturpflanzen bei unterschiedlich versalzten Keimmedien. - Arch.
Acker- Pflanzenbau Bodenk. 24: 241-245, 1980.

*8656 - HEGGESTADT, H.E., HEAGLE, A.S., BENNETT, J.H., KOCH, E.J.: The effects of
photochemical oxidants on the yield of snap beans. - Atmos. Environ. 14:
317-326, 1980.

*8657 - HEIKAL, M.M., AHMED, A.H., SHADDAD, M.A.: Salt tolerance of some oil pro-
ducing plants. - Agricultura (Heverlee) 28: 437-453, 1980.

8658 - HEINE, R.W.: Effect of soil moisture, solar radiation, and dew on wheat ear
water potentials. - N.Zeal. J. agr. Res. 24: 139-140, 1981.

*8659 - HEINE, R.W., RYU, K.S.: Measurement of evapotranspiration under wheat in
Canterbury by monitoring soil moisture fluxes. - N.Zeal. J. Sci. 23: 251-
257, 1980.

8660 - HELAL, H.M., MENGEL, K.: Interaction between light intensity and NaCl salinity and their effects on growth, CO_2 assimilation, and photosynthate con-
version in young broad beans. - Plant Physiol. 67: 999-1002, 1981.

*8661 - HELLEBUST, J.A.: Reactions to water and salt stress in plants. - In:
SPANSWICK, R.M., LUCAS, W.J., DAINTY, J. (ed.): Plant Membrane Transport:
Current Conceptual Issues. Pp. 147-156. Elsevier / North-Holland Biomedical
Press, Amsterdam - New York - Oxford 1980.

*8662 - HELLKVIST, J., HILLERDAL-HAGSTRÖMER, K., MATTSON-DJOS, E.: Field studies of
water relations and photosynthesis in Scots pine using manual techniques. -
Ecol. Bull. (Stockholm) 32 (PERSSON, T. (ed.): Structure and function of
Northern Coniferous Forests - An Ecosystem Study.): 183-204, 1980.

8663 - HELLUM, A.K., BARKER, N.A.: The relationship of lodgepole pine cone age and
seed extractability. - Forest Sci. 27: 62-70, 1981.

8664 - HENSON, I.E.: Changes in abscisic acid content during stomatal closure in
pearl millet (Pennisetum americanum (L.) Leeke). - Plant Sci. Lett. 21:
121-127, 1981.

8665 - HENSON, I.E.: Abscisic acid and after-effects of water stress in pearl millet
(Pennisetum americanum (L.) Leeke). - Plant Sci. Lett. 21: 129-135, 1981.

8666 - HENSON, I.E., MAHALAKSHMI, V., BIDINGER, F.R., ALAGARSWAMY, G.: Genotypic
variation in pearl millet (Pennisetum americanum (L.) Leeke), in the ability
to accumulate abscisic acid in response to water stress. - J. exp. Bot. 32:
899-910, 1981.

8667 - HENSON, I.E., MAHALAKSHMI, V., BIDINGER, F.R., ALAGARSWAMY, G.: Stomatal
responses of pearl millet (Pennisetum americanum (L.) Leeke), genotypes, in
relation to abscisic acid and water stress. - J. exp. Bot. 32: 1211-1221,
1981.

8668 - HENSON, I.E., QUARRIE, S.A.: Abscisic acid accumulation in detached cereal leaves in response to water stress. I. Effects of incubation time and severity of stress. - Z. Pflanzenphysiol. 101: 431-438, 1981.

8669 - HENZELL, E.F.: Forage legumes. - In: BROUGHTON, W.J. (ed.): Nitrogen Fixation. Volume I. Ecology. Pp. 264-289. Clarendon Press, Oxford 1981.

8670 - HERBORN, A., LENZ, F.: Regulierung des Gasdiffusionswiderstands bei veränderter Source-Sink-Beziehung bei Buschbohnenpflanzen (*Phaseolus vulgaris* L.). - Gartenbauwissenschaft 46: 181-186, 1981.

8671 - HERRERO, M.P., JOHNSON, R.R.: Drought stress and its effects on maize reproductive systems. - Crop Sci. 21: 105-110, 1981.

8672 - HETSCH, W., HEILIG, K.-H.: Der Wasserhaushalt von Fichte in Abhängigkeit von Boden und Atmosphäre. - Z. Pflanzenernähr. Bodenk. 144: 317-330, 1981.

8673 - HEUER, B., PLAUT, Z.: Carbon dioxide fixation and ribulose-1,5-bisphosphate carboxylase activity in intact leaves of sugar beet plants exposed to salinity and water stress. - Ann. Bot. 48: 261-268, 1981.

8674 - HEUN, A.-M., GORHAM, J., LÜTTGE, U., WYNJONES, R.G.: Changes of water-relation characteristics and levels of organic cytoplasmic solutes during salinity induced transition of *Mesembryanthemum crystallinum* from C_3-photosynthesis to crassulacean acid metabolism. - Oecologia 50: 66-72, 1981.

8675 - HEYSER, J.W., NABORS, M.W.: Osmotic adjustment of cultured tobacco cells (*Nicotiana tabacum* var. Samsum) grown on sodium chloride. - Plant Physiol. 67: 720-727, 1981.

*8676 - HINCKLEY, T.M., DUHME, F.: Water relations of mediterranean shrub species. - Plant Physiol. 65 (Suppl.): 155, 1980.

8677 - HINCKLEY, T.M., TESKEY, R.O., DUHME, F., RICHTER, H.: Temperate hardwood forests. - In: KOZLOWSKI, T.T. (ed.): Water Deficits and Plant Growth. Volume VI. Woody Plant Communities. Pp. 153-208. Academic Press, New York - San Francisco - London 1981.

*8678 - HITZ, W.D., HANSON, A.D.: Phospholipids as intermediates of betaine synthesis in water-stressed barley leaves. - Plant Physiol. 65 (Suppl.): 8, 1980.

8679 - HITZ, W.D., RHODES, D., HANSON, A.D.: Radiotracer evidence implicating phosphoryl and phosphatidyl bases as intermediates in betaine synthesis by water-stressed barley leaves. - Plant Physiol. 68: 814-822, 1981.

*8680 - HNATIUK, R.J., HOPKINS, A.J.M.: Western Australian species - rich kwongan (sclerophyllous shrubland) affected by drought. - Aust. J. Bot. 28: 573-585, 1980.

8681 - HOBBS, E.H., KROGMAN, K.K.: Sorghum and barley in southern Alberta: Grain yield response to irrigation and fertilizer. - Can. J. Plant Sci. 61: 837-842, 1981.

*8682 - HOCKING, P.J., PATE, J.S., ATKINS, C.A., SHARKEY, P.J.: Diurnal patterns of transport and accumulation of minerals in fruiting plants of *Lupinus angustifolius* L. - Ann. Bot. 42: 1277-1290, 1978.

8683 - HODGSON, L.M.: Photosynthesis of the red alga *Gastroclonium coulteri* (*Rhodophyta*) in response to changes in temperature, light intensity, and desiccation. - J. Phycol. 17: 37-42, 1981.

*8684 - HOFFMANN, A.J., WALKER, M.J.: Growth habits and phenology of drought-deciduous species in an altitudinal gradient. - Can. J. Bot. 58: 1789-1796, 1980.

8685 - HOFFMANN, F.: Modellierung des Wachstums von Pflanzenbeständen, dargestellt am Beispiel der Zuckerrüben. - In: UNGER, K., STÖCKER, G. (ed.): Biophysikalische Ökologie und Ökosystemforschung. Pp. 153-162. Akademie-Verlag, Berlin 1981.

8686 - HOHMANN, B.: Zur mikroskopischen Identifizierung von Amaranthus retroflexus L. in Grünmehlen. - Landwirt. Forsch. 34: 67-72, 1981.

8687 - HOLBO, H.R.: A dew-point hygrometer for field use. - Agr. Meteorol. 24: 117-130, 1981.

*8688 - HÖLLWARTH, M., KULL, U.: Einige ökophysiologische Untersuchungen auf Tenerife (Kanarische Inseln). - Bot. Jahrb. Syst. 100: 518-535, 1979.

*8689 - HONG, S.-G., SUCOFF, E.: Units of freezing of deep supercooled water in woody xylem. - Plant Physiol. 66: 40-45, 1980.

*8690 - HOOPER, A.W.: Estimation of the moisture content of grass from diffuse reflectance measurements at near infrared wavelengths. - J. agr. eng. Res. 25: 355-366, 1980.

8691 - HORVÁTH, I., SEN, S., SZUJKÓ-LACZA, J.: Leaf anatomical reactions of Pimpinella anisum L. to different light intensities. - In: UNGER, K., STÖCKER, G. (ed.): Biophysikalische Ökologie und Ökosystemforschung. Pp. 129-136. Akademie-Verlag, Berlin 1981.

8692 - HROMADKA, T.V., II, GUYMON, G.L.: Subdomain integration model of ground-water flow. - J. Irrig. Drain. Div. ASCE 107: 187-195, 1981.

8693 - HUBAC, C., LE PAGE-DEGIVRY, M.-T.: Évolution de la teneur en acide abscissique au cours de l'assèchement du Cotonnier sous différentes photopériodes. - Physiol. vég. 19: 87-97, 1981.

*8694 - HUBICK, K.T., REID, D.M.: A rapid method for the extraction and analysis of abscisic acid from plant tissue. - Plant Physiol. 65: 523-525, 1980.

8695 - HULSTON, J.R., TAYLOR, C.B., LYON, G.L., STEWART, M.K., COX, M.A.: Environmental isotopes in New Zealand hydrology 2. Standards, measurement techniques, and reporting of measurements for oxygen-18, deuterium, and tritium in water. - N. Zeal. J. agr. Sci. 24: 313-332, 1981.

8696 - HUMBERT, C., GUYOT, M.: Observation épiscopique in vivo des modifications des cellules stomatiques et des cellules épidermiques provoquées par des variations d'éclairement et de teneur en CO$_2$. - Physiol. vég. 19: 167-175, 1981.

*8697 - HUMBERT, C., LOUGUET, P., GUYOT, M.: Modifications ultrastructurales des cellules de garde et mouvements des stomates chez le Pelargonium x hortorum. - Rev. Cytol. Biol. vég. - Bot. 1: 233-257, 1978.

*8698 - HUNER, N.P.A., CARTER, J.V.: Effects of dehydration on the structure of RuBPCASE from cold-hardened and unhardened puma rye. - Plant Physiol. 65 (Suppl.): 154, 1980.

8699 - HUNER, N.P.A., PALTA, J.P., LI, P.H., CARTER, J.V.: Anatomical changes in leaves of Puma rye in response to growth at cold-hardening teperatures. - Bot. Gaz. 142: 55-62, 1981.

8700 - HUNT, L.A., EDGINGTON, L.V.: Dry matter accumulation and distribution in winter wheat grown in a humid continental climate. - Can. J. Bot. 59: 415-420, 1981.

8701 - HUNT, P.G., CAMPBELL, R.E., SOJKA, R.E., PARSONS, J.E.: Flooding-induced soil and plant ethylene accumulation and water status response of field--grown tobacco. - Plant Soil 59: 427-439, 1981.

8702 - HUNT, P.G., WOLLUM, A.G.,II., MATHENY, T.A.: Effects of soil water on *Rhizobium japonicum* infection, nitrogen accumulation, and yield in bragg soybeans. - Agron. J. 73: 501-505, 1981.

8703 - HUNTER, M.N.: Semi-automatic control of soil water in pot culture. - Plant Soil 62: 455-459, 1981.

8704 - HUNTER, M.N., BYTH, D.E., EDWARDS, D.G., ASHER, C.J.: Effects of watering technique and pot size on response of soybean cultivars to applied zinc. - Aust. J. agr. Res. 32: 69-78, 1981.

8705 - HUNTER, M.N., EDWARDS, D.G.: The influence of watering regimen on soil pH and zinc nutrition of soybean cv. Wills grown in soil pot culture. - Aust. J. agr. Res. 32: 871-881, 1981.

*8706 - HÜSKEN, D., ZIMMERMANN, U., SCHULZE, E.-D.: Water relations of leaves of *Tradescantia virginiana*: direct turgor-pressure measurement. - In: SPANSWICK, R.M., LUKAS, W.J., DAINTY, J. (ed.): Plant Membrane Transport: Current Conceptual Issues. Pp. 469-470. Elsevier/ North-Holland Biomedical Press, Amsterdam - New York - Oxford 1980.

8707 - HUTTUNEN, S., HAVAS, P., LAINE, K.: Effects of air pollutants on the wintertime water economy of the Scots pine *Pinus silvestris*. - Holarctic Ecol. 4: 94-101, 1981.

8708 - HUTTUNEN, S., KÄRENLAMPI, L., KOLARI, K.: Changes in osmotic potential and some related physiological variables in needles of polluted Norway spruce (*Picea abies*). - Ann. Bot. Fenn. 18: 63-71, 1981.

8709 - HUZULÁK, J.: Ekologicko-Fyziologická Štúdia Vodného Režimu Lesných Drevín. [Eco-physiological Study of the Forest Woody Species Water Relations.] - Veda, Bratislava 1981. [In Slov, ab: R, E.]

8710 - IDSO, S.B.: Relative rates of evaporative water losses from open and vegetation covered water bodies. - Water Resour. Bull. 17: 46-48, 1981.

8711 - IDSO, S.B., JACKSON, R.D., PINTER, P.J.,Jr., REGINATO, R.J., HATFIELD, J.L.: Normalizing the stress-degree-day parameter for environmental variability. - Agr. Meteorol. 24: 45-55, 1981.

8712 - IDSO, S.B., REGINATO, R.J., JACKSON, R.D., PINTER, P.J.,Jr.: Measuring yield-reducing plant water potential depressions in wheat by infrared thermometry. - Irrig. Sci. 2: 205-212, 1981.

8713 - IDSO, S.B., REGINATO, R.J., JACKSON, R.D., PINTER, P.J.,Jr.: Foliage and air temperatures: Evidence for a dynamic "equivalence point". - Agr. Meteorol. 24: 223-226, 1981.

8714 - IDSO, S.B., REGINATO, R.J., REICOSKY, D.C., HATFIELD, J.L.: Determining soil-induced plant water potential depressions in alfalfa by means of infrared thermometry. - Agron. J. 73: 826-830, 1981.

8715 - IKE, I.F., THURTELL, G.W.: Water relations of cassava: water content, water, osmotic and turgor potential relationships. - Can. J. Bot. 59: 956-964, 1981.

8716 - IKE, I.F., THURTELL, G.W.: Osmotic adjustment in indoor grown cassava in response to water stress. - Physiol. Plant. 52: 257-262, 1981.

*8717 - ILIEV, V.: Fotosintetichna produktivnost na napoyavan sl'nchogled pri razlichni ravnishcha na torene. [Photosynthetic productivity of irrigated sunflower at varying fertilizer levels.] - Rasteniev. Nauki 17 (1): 36-45, 1980. [In Bulg, ab: R, E.]

*8718 - ILOBA, C.: Einfluss der Substratfeuchtigkeit und antibakterieller Antibiotika auf das Vorkommen und auf die Sporenbildung von *Drechslera oryzae* bei routinemässigen Saatgutuntersuchungen. - Z. Pflanzenkrank. Pflanzenschutz 87: 600-606, 1980.

8719 - IL'YASHUK, E.M., BERSHTEIN, B.I., BORSHCH, T.M., LIKHOLAT, D.A.: Vliyanie ekzogennoĭ abstsizovoĭ kisloty na gazoobmen list'ev sakharnoĭ svekly pri ogranichennom vodosnabzhenii. [Effect of exogenic abscisic acid on gas exchange of sugar beet leaves at limited water supply.] - Fiziol. Biokhim. kul't. Rast. 13: 188-193, 1981. [In R, ab: E.]

*8720 - IMAMURA, J., HARADA, H.: Effects of abscisic acid and water stress on the embryo and plantlet formation in anther culture of *Nicotiana tabacum* cv. Samsun. - Z. Pflanzenphysiol. 100: 285-289, 1980.

*8721 - IMOLEHIN, E.D., GROGAN, R.G., DUNIWAY, J.M.: Effect of temperature and moisture tension on growth, sclerotial production, germination, and infection by *Sclerotinia minor*. - Phytopathology 70: 1153-1157, 1980.

8722 - INGESTAD, T., ARONSSON, A., ÅGREN, G.I.: Nutrient flux density model of mineral nutrition in conifer ecosystems. - Studia forest. Suecica 160 (LINDER, S. (ed.): Understanding and Predicting Tree Growth.): 61-71, 1981.

8723 - INNES, P., BLACKWELL, R.D.: The effect of drought on the water use and yield of two spring wheat genotypes. - J. agr. Sci. 96: 603-610, 1981.

8724 - INOUYE, J., HAGIWARA, T.: Effects of some environmental factors on the position of the lowest elongated internode of three floating rice varieties. - Jap. J. trop. Agr. 25: 115-121, 1981.

*8725 - INOUYE, J., MOCHIZUKI, T.: Emergence of crown roots from the elongated culm in several floating rice varieties under submerged conditions. - Jap. J. Trop. Agr. 24: 125-131, 1980.

*8726 - INOUYE, J., MOGAMI, Y.: On the position of the lowest elongated internode of floating rices originated in different countries. - Jap. J. Trop. Agr. 24: 13-17, 1980.

8727 - INUYAMA, S.: Effectiveness of straw mulch for alleviating drought stress of grain sorghum. - Jap. J. Crop Sci. 50: 217-222, 1981.

8728 - ĬORDANOV, Ĭ.G., SHOPOVA, K.: Faktori za povishavane i balansirane na fotosintetichnata i beit'chnata produktivnost na tsarevitsata. [Factors enhancing and balancing the photosynthetic and protein productivity of maize.]- Rasteniev. Nauki 18 (5): 33-43, 1981. [In Bulg, ab: R, E.]

8729 - ISHIHARA, K., HIRASAWA, T., IIDA, O., KIMURA, M.: [Diurnal course of transpiration rate, stomatal aperture, stomatal conductance, xylem water potential and leaf water potential in the rice plants under different growth conditions.] - Jap. J. Crop Sci. 50: 25-37, 1981. [In Jap, ab: E.]

8730 - ISHIKAWA, M., SAKAI, A.: Freezing avoidance mechanisms by supercooling in some *Rhododendron* flower buds with reference to water relations. - Plant Cell Physiol. 22: 953-967, 1981.

8731 - ITIER, B.: Une méthode simple pour la mesure de l'évapotranspiration réelle
à l'échelle de la parcelle. - Agronomie 1: 869-876, 1981.

8732 - JACKSON, R.D., IDSO, S.B., REGINATO, R.J., PINTER, P.J.,Jr.: Canopy temperature as a crop water stress indicator. - Water Resour. Res. 17: 1133-1138,
1981.

8733 - JACOB, F., JÄGER, E.J., OHMANN, E.: Kompendium der Botanik. - Gustav Fischer
Verlag, Stuttgart 1981.

*8734 - JACOBSON, J.S., TROIANO, J., COLAVITO, L.J., HELLER, L.I., McCUNE, D.C.:
Polluted rain and plant growth. - In: TORIBARA, T.Y., MILLER, M.W., MORROW,
P.E. (ed.): Polluted Rain. Environmental Science Research. Volume 17. Pp.
291-299. Plenum Press, New York 1980.

8735 - JACOBSON, M.B., STONER, W.A., RICHARDS, S.P.: Models of plant and soil
processes. - In: MILLER, P.C. (ed.): Resource Use by Chaparral and Matorral.
A Comparison of Vegetative Function in Two Mediterranean Type Ecosystems.
Pp. 287-368. Springer-Verlag, New York - Heidelberg - Berlin 1981.

8736 - JACQUINOT, L., FORGET, M., EDAH, K.A.: Résistance à la transpiration chez le
riz pluvial (Oryza sativa). Étude d'un test de criblage variétal. - Agron.
trop. 36: 247-252, 1981.

*8737 - JAIN, T.C.: Planning for increased crop production under limited water re-
sources in arid and semi-arid conditions. - Ann. Arid Zone 19: 227-230,
1980.

8738 - JAKOBSONS, P., SKJELVÅG, A.O.: Einige Einflusse von Temperatur und Nieder-
schlag auf Struktur und Produktion alter Dauerwiesen in Rendalen, Norwegen.
- Meld. Norg. Landbrukshøgsk. 60 (24): 2-20, 1981.

8739 - JANISTYN, B.: Sukkulenz - Induktion bei Kalanchoe blossfeldiana im Langtag
durch eine lipophile Fraktion aus blühenden Kalanchoe und MS-Identifikation
von Pterosteron. - Z. Naturforsch. 36: 455-458, 1981.

*8740 - JARVIS, P.G.: Stomatal conductance, gaseous exchange and transpiration. -
In: GRACE, J., FORD, E.D., JARVIS, P.G. (ed.): Plants and their Atmospheric
Environment. Pp. 175-204. Blackwell Scientific Publications, Oxford - London
- Edinburgh - Boston - Melbourne 1980.

8741 - JARVIS, P.G.: Plant water relations in models of tree growth. - Studia forest.
Suecica 160 (LINDER, S. (ed.): Understanding and Predicting Tree Growth.):
51-60, 1981.

8742 - JARVIS, P.G.: Production efficiency of coniferous forest in the UK. - In:
JOHNSON, C.B. (ed.): Physiological Processes Limiting Plant Productivity.
Pp. 81-107. Butterworths, London - Boston - Sydney - Wellington - Durban -
Toronto 1981.

8743 - JARVIS, P.G., EDWARDS, W.R.N., TALBOT, H.: Models of plant and crop water
use. - In: ROSE, D.A., CHARLES-EDWARDS, D.A. (ed.): Mathematics and Plant
Physiology. Pp. 151-194. Academic Press, London - New York - Toronto -
Sydney - San Francisco 1981.

8744 - JARVIS, P.G., MANSFIELD, T.A. (ed.): Stomatal Physiology. Society for Experimental Biology Seminar Series Volume 8. - Cambridge University Press,
Cambridge - London - New York - New Rochelle - Melbourne - Sydney 1981.

8745 - JARVIS, P.G., MORISON, J.I.L.: The control of transpiration and photosynthe-
sis by the stomata. - In: JARVIS, P.G., MANSFIELD, T.A. (ed.): Stomatal
Physiology. Pp. 247-279. Cambridge University Press, Cambridge - London -
New York - New Rochelle - Melbourne - Sydney 1981.

8746 - JARVIS, S.C., HOPPER, M.J.: The uptake of sodium by perennial ryegrass and its relationship to potassium supply in flowing solution culture. - Plant Soil 60: 73-83, 1981.

8747 - JAUHAR, P.P.: Cytogenetics and Breeding of Pearl Millet and Related Species. (Progress and Topics in Cytogenetics Volume I.). - Alan R. Liss, Inc., New York 1981.

*8748 - JEANRENAUD, E.: Le comportement hydrique de quelques espèces des pâturages de steppe du littoral de la Mer Noire, Agigea-Constanţa. - Trav. Mus. Hist. natur. "Grigore Antipa" 19: 363-375, 1978.

8749 - JEFFERIES, R.L.: Osmotic adjustment and the response of halophytic plants to salinity. - BioScience 31: 42-46, 1981.

8750 - JENSEN, C.R.: Influence of water and salt stress on water relationships and carbon dioxide exchange of top and roots in beans. - New Phytol. 87: 285-295, 1981.

8751 - JENSEN, C.R.: Influence of soil water stress on wilting and water relations of differently osmotically adjusted wheat plants. - New Phytol. 89: 15-24, 1981.

*8752 - JENSEN, H.E.K.: Virkning af temperatur og luftfugtighed i hvileperioden på udbyttet i vaeksthusroser, Rosa L. [The effect of temperature and air humidity during the rest period on the yield of glasshouse roses, Rosa L.] - Tidsskr. Planteavl 84: 229-236, 1980. [In Dan, ab: E.]

*8753 - JENSEN, K.G., PALTA, J.P.: Plamamembrane surface during osmotically induced volume changes: evidence against membrane folding. - Plant Physiol. 65 (Suppl.): 147, 1980.

8754 - JENSEN, M., HEBER, U., OETTMEIER, W.: Chloroplast membrane damage during freezing: The lipid phase. - Cryobiology 18: 322-335, 1981.

*8755 - JENSÉN, P., KYLIN, A.: Effects of ionic strength and relative humidity on the efflux of K^+ (^{86}Rb) and Ca^{2+} (^{45}Ca) from roots of intact seedlings of cucumber, oat and wheat. - Physiol. Plant. 50: 199-207, 1980.

8756 - JEONG, Y.-H., OTA, Y.: [Physiological studies on photochemical oxidant injury in rice plants III. Relationship between abscisic acid (ABA) and water metabolism in water-stressed rice plants.]- Jap. J. Crop Sci. 50: 566-569, 1981. [In Jap, ab: E.]

8757 - JEWER, P.C., INCOLL, L.D.: Promotion of stomatal opening in detached epidermis of Kalanchoe daigremontiana Hamet et Perr. by natural and synthetic cytokinins. - Planta 153: 317-318, 1981.

8758 - JEWER, P.C., INCOLL, L.D., HOWARTH, G.L.: Stomatal responses in isolated epidermis of the crassulacean acid metabolism plant Kalanchoe daigremontiana Hamet et Perr. - Planta 153: 238-245, 1981.

8759 - JOHNSON, C.B. (ed.): Physiological Processes Limiting Plant Productivity. - Butterworths, London - Boston - Sydney - Wellington - Durban - Toronto 1981.

8760 - JOHNSON, C.R., INGRAM, D.L., BARRETT, J.E.: Effects of irrigation frequency on growth, transpiration, and acclimatization of Ficus benjamina L. - HortScience 16: 80-81, 1981.

8761 - JOHNSON, D.A., RUMBAUGH, M.D., ASAY, K.H.: Plant improvement for semi-arid rangelands: possibilities for drought resistance and nitrogen fixation. - Plant Soil 58: 279-303, 1981. Also in: MONTEITH, J., WEBB, C. (ed.): Soil Water and Nitrogen in Mediterranean-Type Environments. Development in Plant and Soil Sciences. Volume 1. Pp. 279-303. Martinus Nijhof / Dr. W. Junk Publishers, The Hague - Boston - London 1981.

8762 - JOHNSON, J.D.: Two types of ventilated porometers compared on broadleaf and coniferous species. - Plant Physiol. 68: 506-508, 1981.

*8763 - JOHNSON, J.D., FERRELL, W.K.: Leaf conductance as affected by rapid changes in the air-leaf vapor pressure deficit. - Plant Physiol. 65 (Suppl.): 47, 1980.

8764 - JOHNSON, R.C., WITTERS, R.E., CIHA, A.J.: Daily patterns of apparent photosynthesis and evapotranspiration in a developing winter wheat crop. - Agron. J. 73: 414-418, 1981.

8765 - JOHNSON, R.C., WITTERS, R.E., CIHA, A.J.: Apparent photosynthesis, evapotranspiration, and light penetration in two contrasting hard red winter wheat canopies. - Agron. J. 73: 419-422, 1981.

8766 - JOHNSON, R.W., RIDING, R.T.: Structure and ontogeny of the stomatal complex in *Pinus strobus* L. and *Pinus banksiana* Lamb. - Amer. J. Bot. 68: 260-268, 1981.

*8767 - JOHNSON, W.C., DAVIS, R.G.: Recording vapor pressure, relative humidity, or vapor pressure deficit using a lithium chloride dewcel. - Texas agr. Exp. Sta. tech. Rep. 28: 1-31, 1980.

8768 - JONES, C.A.: Effect of drought stress on percentage filled grains in upland rice. - Trop. Agr. (Trinidad) 58: 201-203, 1981.

8769 - JONES, H.G.: The use of stochastic modelling to study the influence of stomatal behaviour on yield-climate relationships. - In: ROSE, D.A., CHARLES -EDWARDS, D.A. (ed.): Mathematics and Plant Physiology. Pp. 231-244. Academic Press, London - New York - Toronto - Sydney - San Francisco 1981.

*8770 - JONES, H.G., HIGGS, K.H.: Resistance to water loss from the mesophyll cell surface in plant leaves. - J. exp. Bot. 31: 545-553, 1980.

*8771 - JONES, H.G., NORTON, T.A.: The role of internal factors in controlling evaporation from intertidal algae. - In: GRACE, J., FORD, E.D., JARVIS, P.G. (ed.): Plants and their Atmospheric Environment. Pp. 231-235. Blackwell Scientific Publications, Oxford - London - Edinburgh - Boston - Melbourne 1980.

8772 - JONES, M.M., TURNER, N.C., OSMOND, C.B.: Mechanisms of drought resistance. - In: PALEG, L.G., ASPINALL, D. (ed.): The Physiology and Biochemistry of Drought Resistance in Plants. Pp. 15-37. Academic Press, Sydney - New York - London - Toronto - San Francisco 1981.

8773 - JORDAN, P.W., NOBEL, P.S.: Seedling establishment of *Ferocactus acanthodes* in relation to drought. - Ecology 62: 901-906, 1981.

*8774 - JØRGENSEN, V.: Vandingsfrekvensens indflydelse på udbytte og vandforbrug i byg. [The effect of irrigation frequency on barley yield and water use.] - Tidsskr. Planteavl 84: 335-341, 1980. [In Dan, ab: E.]

*8775 - JOSHI, A.J., IYENGAR, E.R.R., BHATT, D.C.: Effects of salinity on structure and frequency of stomata in salt marsh halophytes. - Geobios 7: 210-213, 1980.

8776 - JOSHI, G.V., WAGHMODE, A.P.: Photosynthesis in mangroves. - Ind. J. Bot. 4: 15-19, 1981.

8777 - JURY, W.A., LETEY, J.,Jr., STOLZY, L.H.: Flow of water and energy under desert conditions. - In: EVANS, D.D., THAMES, J.L. (ed.): Water in Desert Ecosystems. Pp. 92-113. Dowden, Hutchinson & Ross, Inc., Strodsburg 1981.

8778 - KABAKI, N., TAJIMA, K.: Effect of chilling on the water balance of rice seedlings. - Jap. J. Crop Sci. 50: 489-494, 1981.

*8779 - **KAFKAFI, U., BAR-YOSEF, B.**: Trickle irrigation and fertilization of tomatoes in highly calcareous soils. - Agron. J. 72: 893-897, 1980.

8780 - **KAISER, W.M., HEBER, U.**: Photosynthesis under osmotic stress. Effect of high solute concentrations on the permeability properties of the chloroplast envelope and on activity of stroma enzymes. - Planta 153: 423-429, 1981.

8781 - **KAISER, W.M., KAISER, G., PRACHUAB, P.K., WILDMAN, S.G., HEBER, U.**: Photosynthesis under osmotic stress. Inhibition of photosynthesis of intact chloroplasts, protoplasts, and leaf slices at high osmotic potentials. - Planta 153: 416-422, 1981.

8782 - **KAISER, W.M., KAISER, G., SCHÖNER, S., NEIMANIS, S.**: Photosynthesis under osmotic stress. Differential recovery of photosynthetic activities of stroma enzymes, intact chloroplasts, protoplasts, and leaf slices after exposure to high solute concentrations. - Planta 153: 430-435, 1981.

8783 - **KAISER, W.M., STEPPER, W., URBACH, W.**: Photosynthesis of isolated chloroplasts and protoplasts under osmotic stress. Reversible swelling of chloroplasts by hypotonic treatment and its effect on photosynthesis. - Planta 151: 375-380, 1981.

8784 - **KAKU, S., IWAYA, M., JEON, K.B.**: Supercooling ability, water content and hardiness of *Rhododendron* flower buds during cold acclimation and deacclimation. - Plant Cell Physiol. 22: 1561-1569, 1981.

*8785 - **KANEVS'KYĬ, V.O., MIROLYUBOV, O.V., TKACHENKO, V.S., SHELYAG-SOSONKO, Yu.R.**: Zastosuvannya metodu aktyvnoï radiolokatsiï dlya vyvchennya vplyvu melioratsiï na vodnyĭ rezhym roslyn. [Application of the active radiolocation method for studying the effect of reclamation on water regime of plants.] - Ukr. bot. Zh. 37 (6): 83-86, 1980. [In Ukr, ab: R, E.]

8786 - **KAO, C.H.**: Senescence of rice leaves VI. Comparative study of the metabolic changes of senescing turgid and water-stressed excised leaves. - Plant Cell Physiol. 22: 683-688, 1981.

8787 - **KAPPEN, L.**: Ecological significance of resistance to high temperature. - In: LANGE, O.L., NOBEL, P.S., OSMOND, C.B., ZIEGLER, H. (ed.): Physiological Plant Ecology I. Responses to the Physical Environment. Pp. 439-474. Springer-Verlag, Berlin - Heidelberg - New York 1981.

8788 - **KAPPEN, L., FRIEDMANN, E.I., GARTY, J.**: Ecophysiology of lichens in the dry valleys of Southern Victoria Land, Antarctica I. Microclimate of the cryptoendolithic lichen habitat. - Flora 171: 216-235, 1981.

8789 - **KARA, O., PULLI, S.**: Rehumaissin typpilannoituksesta ja sadetuksesta. [Nitrogen fertilization and irrigation of silage maize in Finland.] - J. sci. agr. Soc. Finland 53: 64-74, 1981. [In Fin, ab: E.]

*8790 - **KARADGE, B.A., JOSHI, G.V.**: Carbon assimilation & crassulacean acid metabolism in *Portulaca oleracea* Linn. - Ind. J. exp. Biol. 18: 631-634, 1980.

*8791 - **KARLSTRÖM, P.-O.**: Epidermal leaf structures in species of Asystasieae, Pseuderanthemeae, Graptophylleae and Odontonemeae (Acanthaceae). - Bot. Notiser 133: 1-16, 1980.

8792 - **KARMANOV, V.G., ODUMANOVA-DUNAEVA, G.A.**: O roli fotosinteza v preryvanii svetom temnovykh protsessov fotoperiodicheskoĭ reaktsii rasteniĭ. [On the role of photosynthesis in the light interruption of dark processes of photoperiodical reaction in plants.]- Bot. Zh. 66: 674-683, 1981. [In R, ab: E.]

8793 - **KARMANOV, V.G., ODUMANOVA-DUNAEVA, G.A., SOLOV'EV, E.V.:** Sopryazhennye izme-
neniya fotosinteza, vodnogo i teplovogo rezhimov rasteniĭ v zavisimosti ot
faktorov vneshneĭ sredy. [Correlation of photosynthesis, water- and heat
regimes of plants depending on the environmental factors.] - Bot. Zh. 66:
502-514, 1981. [In R, ab: E.]

8794 - **KASAMO, K.:** Effect of abscisic acid on the K^+ efflux and membrane potential
of *Nicotiana tabacum* L. leaf cells. - Plant Cell Physiol. 22: 1257-1267,
1981.

8795 - **KASAMO, K., SHIMOMURA, T.:** Effect of cold osmotic shock on K^+ efflux from
Nicotiana tabacum leaf discs induced by abscisic acid and ionophores. -
Plant Cell Physiol. 22: 939-951, 1981.

8796 - **KASTNER, W.W.,Jr., GOEBEL, C.J., MAGUIRE, J.D.:** Effects of a wet-dry seed
treatment on the germination and root elongation of "Whitmar" beardles wheat-
grass under various water potentials. - J. Range Manage. 34: 305-307, 1981.

8797 - **KATAOKA, K., TAKAGAKI, E., IKEGAMI, Y.:** [Relationships between yield stabi-
lity and phenotypic plasticity of characters related to photosynthesis in
rice varieties.] - Jap. J. Breed. 31: 65-71, 1981. [In Jap, ab: E.]

*8798 - **KAUFMANN, M.R.:** Automatic determination of conductance and transpiration of
forest trees. - Plant Physiol. 65 (Suppl.): 49, 1980.

8799 - **KAUFMANN, M.R.:** Development of water stress in plants. - HortScience 16:
34-36, 1981.

8800 - **KAUFMANN, M.R.:** Water relations during drought. - In: PALEG, L.G., ASPINALL,
D. (ed.): The Physiology and Biochemistry of Drought Resistance in Plants.
Pp. 55-70. Academic Press, Sydney - New York - Toronto - London - San Fran-
cisco 1981.

8801 - **KAUFMANN, M.R.:** Automatic determination of conductance, transpiration, and
environmental conditions in forest trees. - Forest Sci. 27: 817-827, 1981.

8802 - **KAUFMANN, M.R., TROENDLE, C.A.:** The relationship of leaf area and foliage
biomass to sapwood conducting area in four subalpine forest tree species. -
Forest Sci. 27: 477-482, 1981.

8803 - **KAUL, R., REISENER, H.J.:** Effects of low temperature on potential net photosynthesis in two field-grown winter cereals. - Ann. Bot. 47: 335-338, 1981.

8804 - **KAWASE, M.:** Anatomical and morphological adaptation of plants to waterlogging.
- HortScience 16: 30-34, 1981.

*8805 - **K"DREV, T.G., PETROVA, L.I.:** Vliyanie na nyakoi rastezhni regulatori v"rchu
vodniya rezhim na zakharno tsveklo s razlichna ploidnost'. [Influence of
some growth regulators on the water balance of sugar beet with varying ploi-
dity.] - Fiziol. Rast. (Sofia) 4 (4): 71-77, 1978. [In Bulg, ab: E, R.]

8806 - **KEEFE, P.D., MOORE, K.G.:** Freeze desiccation: a second mechanism for the
survival of hydrated lettuce (*Lactuca sativa* L.) seed at sub-zero temperatures. - Ann. Bot. 47: 635-645, 1981.

8807 - **KEIM, D.L., KRONSTAD, W.E.:** Drought response of winter wheat cultivars grown
under field stress conditions. - Crop Sci. 21: 11-15, 1981.

*8808 - **KEITH, A.D., MASTRO, A., SNIPES, W.:** Diffusion relationships between cellular
plasma membrane and cytoplasm. - In: LYONS, J.M., GRAHAM, D., RAISON, J.K.
(ed.): Low Temperature Stress in Crop Plants. The Role of the Membrane. Pp.
437-451. Academic Press, New York - San Francisco - London 1979.

*8809 - KELLIHER, F.M., TAUER, C.G.: Stomatal resistance and growth of drought
-stressed eastern cottonwood from a wet and dry site. - Silvae Genet. 29:
166-171, 1980.

*8810 - KEMP, P.R., CUNNINGHAM, G.L.: Irradiance, temperature and salinity effects
on growth, leaf anatomy and photosynthesis of *Distichlis spicata* (L.) Greene.
- WRRI (New Mexico Water Resources Research Institute) Rep. 121: I-VI, 1-32,
1980.

8811 - KEMP, P.R., CUNNINGHAM, G.L.: Light, temperature and salinity effects on
growth, leaf anatomy and photosynthesis of *Distichlis spicata* (L.) Greene. -
Amer. J. Bot. 68: 507-516, 1981.

8812 - KENG, J.C.W., SCOTT, T.W., LUGO-LÓPEZ, M.A.: Fertilizer for sweet pepper
under drip irrigation in an oxisol in northwestern Puerto-Rico. - J. agr.
Univ. Puerto Rico 65: 123-128, 1981.

8813 - KENNEDY, R.A., JOHNSON, D.: Changes in photosynthetic characteristics during
leaf development in apple. - Photosynthesis Res. 2: 213-223, 1981.

8814 - KENYON, W.H., HOLADAY, A.S., BLACK, C.C.: Diurnal changes in metabolite
levels and Crassulacean acid metabolism in *Kalanchoë daigremontiana* leaves.
- Plant Physiol. 68: 1002-1007, 1981.

8815 - KESSLY, D.S., BROWN, A.D.: Salt relations of *Dunaliella*. Transitional changes in glycerol content and oxygen exchange reactions on water stress. -
Arch. Microbiol. 129: 154-159, 1981.

8816 - KHAN, A.A., KARSSEN, C.M.: Changes during light and dark osmotic treatment
independently modulating germination and ribonucleic acid synthesis in
Chenopodium-bonus-henricus seeds. - Physiol. Plant. 51: 269-276, 1981.

*8817 - KHAN"MOVA, T., POLONI, E.: Vliyanie na vlazhnostta na tsarevichnoto z"rno
v"rkhu sterilizirashchto deĭstvie na gama-l"chite. [The influence of maize
kernel moisture on the sterilizing effect of gamma rays.] - Rasteniev. Nauki
17 (6): 3-11, 1980. [In Bulg, ab: E, R.]

*8818 - KHANNA-CHOPRA, R., CHATURVERDI, G.S., AGGARWAL, P.K., SINHA, S.K.: Effect of
potassium on growth and nitrate reductase during water stress and recovery
in maize. - Physiol. Plant. 49: 495-500, 1980.

*8819 - KHARE, P.K.: Epidermal structure and development of stomata and trichomes in
Hemionitis arifolia (Burm.) Moore. - Geophytology 10: 140-145, 1980.

8820 - KHVOSTOVA, I.V.: Vliyanie razlichnoĭ vodoobespechennosti kleshcheviny na
kachestvennyĭ sostav belkovo-lipidnogo kompleksa semyan. [Effect of different water supply of castor-oil plant on qualitative composition of protein
-lipid complex of seeds.] - Fiziol. Biokhim. kul't. Rast. 13: 29-34, 1981.
[In R., ab: E.]

*8821 - KIM, J.H., LEE-STADELMANN, O.Y.: Rapid recovery of *Phaseolus vulgaris* leaves
from water stress and differential wilting resistance of basal and trifoliate
leaves. - Plant Physiol. 65 (Suppl.): 8, 1980.

8822 - KIMES, D.S., MARKHAM, B.L., TUCKER, C.J., McMURTREY, J.E.,III.: Temporal
relationships between spectral response and agronomic variables of a corn
canopy. - Remote Sensing Environ. 11: 401-411, 1981.

8823 - KIMMERER, T.W., KOZLOWSKI, T.T.: Stomatal conductance and sulfur uptake of
five clones of *Populus tremuloides* exposed to sulfur dioxide. - Plant Physiol.
67: 990-995, 1981.

8824 - **KINCAID, D.T., LYONS, E.E.:** Winter water relations of red spruce on Mount Monadnock, New Hampshire. - Ecology 62: 1155-1161, 1981.

*8825 - **KINGSOLVER, J.G.:** Thermal and hydric aspects of environmental heterogeneity in the pitcher plant mosquito. - Ecol. Monogr. 49: 357-376, 1979.

8826 - **KIRKHAM, M.B.:** Effects of steroids on water relations and ion uptake of wheat plants. - Biochem. Physiol. Pflanz. 176: 524-534, 1981.

8827 - **KIRKHAM, M.B., HOLDER, P.L.:** Water, osmotic, and turgor potentials of kinetin -treated callus. - HortScience 16: 306-307, 1981.

8828 - **KIRKLAND, K.J., KEYS, C.H.:** The effect of snow trapping and cropping sequence on moisture conservation and utilization in west-central Saskatchewan. - Can. J. Plant Sci. 61: 241-246, 1981.

8829 - **KIRST, G.O.:** Photosynthesis and respiration of *Griffithsia monilis* (Rhodophyceae): Effect of light, salinity, and oxygen. - Planta 151: 281-288, 1981.

*8830 - **KIRST, G.O., BISSON, M.A.:** Osmotic adaptation in marine algae. - In: SPANSWICK, R.M., LUCAS, W.J., DAINTY, J. (ed.): Plant Membrane Transport: Current Conceptual Issues. Pp. 485-486. Elsevier / North-Holland Biomedical Press, Amsterdam - New York - Oxford 1980.

8831 - **KISHITANI, S., TSUNODA, S.:** Physiological aspects of domestication in diploid wheat. - Euphytica 30: 247-252, 1981.

*8832 - **KISLYUK, I.M., GORBAN', I.S.:** Vliyanie temperatury vyrashchivaniya na temperaturnuyu zavisimost' fotosinteza i teploustoĭchivost' nekotorykh funktsiĭ kletok list'ev *Tradescantia albiflora*. [The influence of growth temperature on the temperature dependence of photosynthesis and thermal resistance of some functions of the cells in *Tradescantia albiflora*.]- Bot. Zh. 65: 1383-1391, 1980. [In R, ab: E.]

*8833 - **KLEIN, A.O.:** Rhythms and timing - a review. - In: De GREEF, J. (ed.): Photoreceptors and Plant Development. Pp. 579-598. Antwerpen University Press, Antwerpen 1980.

8834 - **KLEIN, M., DAMAŠKA, J., FÜRST, Z.:** Výnosová reakce obilovin na hydrotermické podmínky přírodních stanovišť. [The yield response of cereals on the hydrothermic conditions of natural sites.] - Rost. Výroba (Praha) 27: 511-516, 1981. [In Czech, ab: R, E, G.]

8835 - **KLOCKARE, R., FALK, S.O.:** Influence of O_2- and CO_2-concentrations on oscillations in the transpiration rate from oat plants in darkness. - Physiol. Plant. 52: 83-88, 1981.

8836 - **KLOSSON, R.J., KRAUSE, G.H.:** Freezing injury in cold-acclimated and unhardened spinach leaves I. Photosynthetic reactions of thylakoids isolated from frost -damaged leaves. - Planta 151: 339-346, 1981.

*8837 - **KLUGE, M., BÖCHER, M., JUNGNICKEL, G.:** Metabolic control of Crassulacean Acid Metabolism: evidence for diurnally changing sensitivity against inhibition by malate of PEP-carboxylase in *Kalanchoë tubiflora* Hamet. - Z. Pflanzenphysiol. 97: 197-204, 1980.

8838 - **KLUGE, M., BÖHLKE, C., QUEIROZ, O.:** Crassulacean acid metabolism (CAM) in *Kalanchoë*: Changes in intercellular CO_2 concentration during a normal CAM cycle and during cycles in continuous light or darkness. - Planta 152: 87-92, 1981.

8839 - KNAPP, A.K., SMITH, W.K.: Water relations and succession in subalpine conifers in southeastern Wyoming. - Bot. Gaz. 142: 502-511, 1981.

8840 - KNIGHT, D.H., FAHEY, T.J., RUNNING, S.W., HARRISON, A.T., WALLACE, L.L.:
Transpiration from 100-yr-old lodgepole pine forests estimated with whole
-tree potometers. - Ecology 62: 717-726, 1981.

8841 - KNIGHT, D.H., FAHEY, T.J., RUNNING, S.W., HARRISON, A.T., WALLACE, L.L.:
Whole-tree potometers estimate pine forest transpiration. - BioScience
31: 242-243, 1981.

8842 - KNITTEL, H., BEHRENDT, S., SCHOTT, P.E.: Einfluss eines Wachstumregulators
auf Wachstum und Ertrag von Wintergerste bei unterschiedlichen Lagerdruck. -
Z. Acker- Pflanzenbau 150: 50-61, 1981.

8843 - KOBATA, T., TAKAMI, S.: [Maintenance of the grain growth in rice (Oryza
sativa L.) subject to water stress during the early grain filling.] - Jap.
J. Crop Sci. 50: 536-545, 1981. [In Jap, ab: E.]

8844 - KOBAYASHI, K., FUCHIGAMI, L.H., BRAINERD, K.E.: Ethylene and ethane production and electrolyte leakage of water-stressed 'Pixy' plum leaves. -
HortScience 16: 57-59, 1981.

8845 - KOLBASINA, Ê.I., ZHIL'TSOVA, V.V.: Vozdeĭstvie zatopleniya na rasteniya
ozimoĭ pshenitsy i rzhi. [Effect of flooding on winter wheat and rye
plants.] - Fiziol. Biokhim. kul't. Rast. 13: 600-605, 1981. [In R, ab: E.]

*8846 - KONDO, N., MARUTA, I., SUGAHARA, K.: Abscisic acid-dependent changes in
transpiration rate with SO_2 fumigation and the effects of sulfite and pH on
stomatal aperture. - Res. Rep. nat. Inst. environ. Studies 11 (Studies on
the Effects of Air Pollutants on Plants and Mechanisms of Phytotoxicity.):
127-136, 1980.

8847 - KORBAN, S.S., COYNE, D.P., WEIHING, J.L.: Rate of water uptake and sites
of water entry in seeds of different cultivars of dry bean. - HortScience
16: 545-546, 1981.

8848 - KORMANOVSKIĬ, A.Ya., IVANOV, I.I., SOKOLOV, Z.M., MATORIN, D.N., VENEDIKTOV,
P.S.: Selektivnaya okislitel'naya destruktsiya pigmentov membran izoliro-
vannykh khloroplastov i ikh funktsional'noe sostoyanie. [Selective oxidative
destruction of pigments of isolated chloroplast membranes and their functio-
nal state.] - Biol. Nauki 1981(7): 37-42, 1981. [In R.]

8849 - KÖRNER, C.: Stomatal behaviour and water potential in apricot with symptoms
of wilt disease. - Angew. Bot. 55: 469-476, 1981.

*8850 - KÖRNER, C., MAYR, R.: Stomatal behaviour in alpine plant communities between
600 and 2600 metres above sea level. - In: GRACE, J., FORD, E.D., JARVIS,
P.G. (ed.): Plants and Their Atmospheric Environment. Pp. 205-218. Blackwell
Scientific Publications, Oxford - London - Edinburgh - Boston - Melbourne
1980.

*8851 - KÖRNER, C., WIESER, G., GUGGENBERGER, H.: Der Wasserhaushalt eines alpinen
Rasens in den Zentralalpen. - In: FRANZ, H. (ed.): Untersuchungen an Alpinen
Böden in den Hohen Tauern 1974-1978 Stoffdynamik und Wasserhaushalt. Pp.
243-264. Universitätsverlag Wagner, Innsbruck 1980.

8852 - KOUCHKOVSKY, Y.,de, HARAUX, F.: ^2H_2O effect on the electron and proton flow
in isolated chloroplasts. An indication for lateral heterogeneity of membrane
pH. - Biochem. biophys. Res. Commun. 99: 205-212, 1981.

8853 - KOVACHEVA, I., DIMITROV, D.A.: Prouchvane v"rkhu napoyavaneto i g"stotata na
poseva pri sortove soya s razlichen vegetatsionen period. I. Vliyanie v"rkhu
nyakoi biologichni, morfologichni i reproduktivni priznatsi na rasteniyata.

[A study on the irrigation and the plant density of soybean cultivars with different vegetation periods. I. Effect on some biological, morphological and reproductive plant characters.] - Rasteniev. Nauki 18 (6): 23-30, 1981. [In Bulg, ab: R, E.]

✲8854 - KOZLOWSKI, T.T.: Impacts of air pollution on forest ecosystems. - BioScience 30: 88-93, 1980.

8855 - KOZLOWSKI, T.T. (ed.): Water Deficits and Plant Growth. Volume VI. Woody Plant Communities. - Academic Press, New York - San Francisco - London 1981.

✲8856 - KRAFTI, G., SADOVSKI, A., KHRISTOV, I., KOLEV, B.: Prognozirane i upravlenie na rezhima na napoyavane. [Prognosis and management of the irrigation regime.] - Rasteniev. Nauki 17 (4): 135-143, 1980. [In Bulg, ab: E, R.]

✲8857 - KRAMER, P.J., KOZLOWSKI, T.T.: Physiology of Woody Plants. - Academic Press, New York - San Francisco - London 1979.

8858 - KRANTZ, B.A.: Water conservation, management, and utilization in semiarid lands. - In: MANASSAH, J.T., BRISKEY, E.J. (ed.): Advances in Food Producing Systems for Arid and Semiarid Lands. Part A. Pp. 339-378. Academic Press, New York - San Francisco - London 1981.

✲8859 - KRAUSE, C.R.: Scanning electron microscopic detection of injury to hybrid poplar leaves induced by ambient air pollution. - Scanning Electron Microscopy 1980: 591-594, 1980.

8860 - KRIEDEMANN, P.E., BARRS, H.D.: Citrus orchards. - In: KOZLOWSKI, T.T. (ed.): Water Deficits and Plant Growth. Volume VI. Woody Plant Communities. Pp. 325-417. Academic Press, New York - San Francisco - London 1981.

8861 - KRIEDEMANN, P.E., DOWNTON, W.J.S.: Photosynthesis. - In: PALEG, L.G., ASPINALL, D. (ed.): The Physiology and Biochemistry of Drought Resistance in Plants. Pp. 283-314. Academic Press, Sydney - New York - London - Toronto - San Francisco 1981.

8862 - KROLL, R.G., ANAGNOSTOPOULOS, G.D.: Potassium fluxes on hyperosmotic shock and the effect of phenol and bronopol (2-bromo-2-nitropropan-1,3-diol) on deplasmolysis of Pseudomonas aeruginosa. - J. appl. Bacteriol. 51: 313-323, 1981.

✲8863 - KRSTIĆ, B., STANKOVIĆ, Ž.: Usvajanje i metabolizam ugljenika. [Carbon assimilation and metabolism.] - In: BELIĆ, J. (ed.): Fiziologija Kukuruza. Pp. 43-64. Serb. Acad. Sci. Arts, Deograd 1980. [In Serb, ab: E.]

8864 - KRÜGER, W., BENKENSTEIN, H.: Einfluss verschiedener Bodenfeuchtestufen auf die Ausnutzung von Mineraldüngerstickstoff und Ertrag von Hafer. - Arch. Acker- Pflanzenbau Bodenk. 25: 51-59, 1981.

8865 - KRUPYANSKIĬ, Yu.F., GAUBMAN, E.Ė., SHAĬTAN, K.V., GOL'DANSKIĬ, V.I., RUBIN, A.B., SUZDALEV, I.P., FROLOV, E.N., SHVEDCHIKOV, A.P., SHCHUKIN, N.F.: Issledovanie dinamiki khromatoforov s pomoshch'yu rėleevskogo rasseyaniya mėssbauėrovskogo izlucheniya. [Investigation of the dynamics of chromatophores by rayleigh scattering of Mössbauer radiation.] - Mol. Biol. (Moskva) 15: 1109-1122, 1981. [In R, ab: E.]

8866 - KUBICHEK, S.A.: Obrazovanie khloroplastov v temnote v zamykayushchikh kletkakh ust'its tradeskantsii virginskoĭ. [Formation of chloroplasts in stomatal guard cells of Tradescantia virginica L. in darkness.]- Fiziol. Rast. 28: 767-773, 1981. [In R, ab: E.]

8867 - KUEH, J.S.H., BRIGHT, S.W.J.: Proline accumulation in a barley mutant resistant to trans-4-hydroxy-L-proline. - Planta 153: 166-171, 1981.

*8868 - KUIPER, F., DE BOER, A.H.: Active and passive transport of ions into the root xylem as a function of the volume flow. - In: SPANSWICK, R.M., LUCAS, W.J., DAINTY, J. (ed.): Plant Membrane Transport: Current Conceptual Issues. Pp. 413-414. Elsevier / North-Holland Biomedical Press, Amsterdam - New York - Oxford 1980.

8869 - KUMAR, D., CHAUHAN, R.P.S., SINGH, R.V.: Salt tolerance of some induced mutants of 'HD 2009' wheat. - Ind. J. agr. Sci. 51: 475-479, 1981.

*8870 - KUPKANCHANAKUL, T., VERGARA, B.S.: Nitrogen response of rice grown in medium deep water. - Thai J. agr. Sci. 13: 213-225, 1980.

8871 - KURAISHI, S., HASHIMOTO, Y., SHIRAISHI, M.: Latent periods of cytokinin-induced stomatal opening in the sunflower leaf. - Plant Cell Physiol. 22: 911-916, 1981.

*8872 - KURAISHI, S., HASHIMOTO, Y., TAKIUCHI, M., ANDRADE LIMA, D.,de: Stomatal aperture of the plants grown in a Brasilian desert, Caatinga. - Environ. Control Biol. 16: 113-118, 1978.

*8873 - KURAISHI, S., NITO, N.: The maximum leaf surface temperatures of the higher plants observed in the Inland Sea area. - Bot. Mag. (Tokyo) 93: 209-220, 1980.

*8874 - KUSHNIRENKO, M.D., KRYUKOVA, E.V., PECHERSKAYA, S.N.: Reaktsiya rasteniĭ grusti na obezvozhivanie i regidratatsiyu. [Reaction of pear plants to dehydration and rehydration.] - Izv. Akad. Nauk Mold. SSR, Ser. biol. khim. Nauk 1980(1): 52-60, 1980. [In R.]

8875 - KUZMANOFF, K.M., EVANS, M.L.: Kinetics of adaptation to osmotic stress in lentil (Lens culinaris Med.) roots. - Plant Physiol. 68: 244-247, 1981.

8876 - KUZNETSOVA, G.A., KUZNETSOVA, M.G., GRINEVA, G.M.: Ob osobennostyakh vodnogo obmena i anatomo-morfologicheskoĭ struktury rasteniĭ kukuruzy pri zatoplenii. [Some features of water relations and anatomical and morphological structure of maize plants under flooding.] - Fiziol. Rast. 28: 340-348, 1981. [In R, ab: E.]

8877 - KYRIAKOPOULOS, E., RICHTER, H.:Pressure - volume curves and drought injury. - Physiol. Plant. 52: 124-128, 1981.

*8878 - LABANAUSKAS, C.K., BINGHAM, F.T., CERDA, A.: Free and protein amino acids, and nutrient concentrations in wheat grain as affected by phosphorus nutrition at various salinity levels. - Plant Soil 49: 581-593, 1978.

8879 - LABANAUSKAS, C.K., SHOUSE, P., STOLZY, L.H.: Effects of water stress at various growth stages on seed yield and nutrient concentrations of field-grown cowpeas. - Soil Sci. 131: 249-256, 1981.

8880 - LABANAUSKAS, C.K., STOLZY, L.H., HANDY, M.F.: Protein and free amino acids in wheat grain as affected by soil types and salinity levels in irrigation water. - Plant Soil 59: 299-316, 1981.

*8881 - LABOURIAU, L.G.: Effects of deuterium oxide on the lower temperature limit of seed germination. - J. therm. Biol. 5: 113-117, 1980.

8882 - LADIGES, P.Y., FOORD, P.C., WILLIS, R.J.: Salinity and waterlogging tolerance of some populations of Melaleuca ericifolia Smith. - Aust. J. Ecol. 6: 203-215, 1981.

8883 - LAHAV, E., KALMAR, D.: Shortening the irrigation interval as a means of saving water in a banana plantation. - Aust. J. agr. Res. 32: 465-477, 1981.

*8884 - LAING, I., UTTING, S.D.: The influence of salinity on the production of two commercially important unicellular marine algae. - Aquaculture 21: 79-86, 1980.

*8885 - LAL, B., AMBASHT, R.S.: Growth of *Chrozophora rottleri* A. Juss in relation to different watering levels. - Ind. J. Ecol. 5: 172-180, 1978.

8886 - LAMBERS, H., BLACQUIÈRE, T., STUIVER, B.(C.E.E.): Interactions between osmo-regulation and the alternative respiratory pathway in *Plantago coronopus* as affected by salinity. - Physiol. Plant. 51: 63-68, 1981.

*8887 - LAMBERS, H., POSTHUMUS, F.: The effect of light intensity and relative humi-dity on growth rate and root respiration of *Plantago lanceolata* and *Zea mays*. - J. exp. Bot. 31: 1621-1630, 1980.

8888 - LAMM, F.R., GREGORY, J.M., CENGIZ, H.S.: The evaluation of a leaf-water potential function for corn. - Trans. ASAE 24: 1172-1176, 1981.

8889 - LAMONT, B.: Availability of water and inorganic nutrients in the persistent leaf bases of the grasstree *Kingia australis*, and uptake and translocation of labelled phosphate by the embedded aerial roots. - Physiol. Plant. 52: 181-186, 1981.

8890 - LANGE, O.L., MATTHES, U.: Moisture-dependent CO_2 exchange of lichens. - Photosynthetica 15: 555-574, 1981.

8891 - LANGE, O.L., NOBEL, P.S., OSMOND, C.B., ZIEGLER, H. (ed.): Physiological Plant Ecology I. Responses to the Physical Environment. (Encyclopedia of Plant Physiology, New Series. Volume 12 A). - Springer-Verlag, Berlin - Heidelberg - New York 1981.

8892 - LANGE, O.L., TENHUNEN, J.D.: Moisture content and CO_2 exchange of lichens. II. Depression of net photosynthesis in *Ramalina maciformis* at high water content is caused by increased thallus carbon dioxide diffusion resistance. - Oecologia 51: 426-429, 1981.

*8893 - LANGELLIER, P.: Determination du regime hydrique d'une culture de canne a sucre de milieu, en periode de maturation, dans le Nord de la Cote d'Ivoire. - Agron. trop. 35: 232-239, 1980.

8894 - LARCHER, W.: Effects of low temperature stress and frost injury on plant productivity. - In: JOHNSON, C.B. (ed.): Physiological Processes Limiting Plant Productivity. Pp. 253-269. Butterworths, London - Boston - Sydney - Wellington - Durban - Toronto 1981.

8895 - LARCHER, W., BAUER, H.: Ecological significance of resistance to low tempe-rature. - In: LANGE, O.L., NOBEL, P.S., OSMOND, C.B., ZIEGLER, H. (ed.): Physiological Plant Ecology I. Responses to the Physical Environment. Pp. 403-437. Springer-Verlag, Berlin - Heidelberg - New York 1981.

8896 - LARSON, E.M., HESKETH, J.D., WOOLLEY, J.T., PETERS, D.B.: Seasonal variations in apparent photosynthesis among plant stands of different soybean cultivars. - Photosynthesis Res. 2: 3-20, 1981.

8897 - LARSSON, S.: Influence of intercepted water on transpiration and evaporation of *Salix*. - Agr. Meteorol. 23: 331-338, 1981.

8898 - LASCÉVE, G., COUCHAT, P.: Intervention de l'oxygène atmosphérique sur l'absorption d'eau chez *Helianthus annuus*. - Physiol. Plant. 52: 47-52, 1981.

8899 - LASSOIE, J.P., SALO, D.J.: Physiological response of large Douglas-fir to natural and induced soil water deficits. - Can. J. Forest Res. 11: 139-144, 1981.

8900 - LAVENDER, D.P.: Environment and shoot growth of woody plants. - Forest Res. Lab., Oregon State Univ. Res. Paper 45: 1-47, 1981.

8901 - LAWLOR, D.W., DAY, W., JOHNSTON, A.E., LEGG, B.J., PARKINSON, K.J.: Growth of spring barley under drought: crop development, photosynthesis, dry-matter accumulation and nutrient content. - J. agr. Sci. 96: 167-186, 1981.

8902 - LAWLOR, D.W., PEARLMAN, J.G.: Compartmental modelling of photorespiration and carbon metabolism of water stressed leaves. - Plant Cell Environ. 4: 37-52, 1981.

8903 - LAYNE, R.E.C., TAN, C.S., FULTON, J.M.: Effect of irrigation and tree density on peach production. - J. Amer. Soc. hort. Sci. 106: 151-156, 1981.

8904 - LAYZELL, D.B., PATE, J.S., ATKINS, C.A., CANVIN, D.T.: Partitioning of carbon and nitrogen and the nutrition of root and shoot apex in a nodulated legume. - Plant Physiol. 67: 30-36, 1981.

8905 - LEE, D.R.: Synchronous pressure-potential changes in the phloem of *Fraxinus americana* L. - Planta 151: 304-308, 1981.

8906 - LEE-STADELMANN, O.Y., HULME, D.A., STADELMANN, E.J.: The apparent viscosity of the protoplasm of subepidermal stem basis cells of *Pisum sativum*. Relation to aging, drought tolerance and water stress. - Physiol. Plant. 52: 65-76, 1981.

*8907 - LEE-STADELMANN, O.Y., PALTA, J.P., LI, P.H.: Leaf elongation during recovery from water stress of *Solanum tuberosum* under controlled environment. - Plant Physiol. 65 (Suppl.):6, 1980.

*8908 - LEMAIRE, G., ROBERGE, G.: Mise au point d'un dispositif expérimental permettant de suivre la consommation hydrique d'une prairie au cours de sa croissance. - Ann. agron. 31: 455-471, 1980.

8909 - LENZ, F., BROUGHTON, W.J.: Growth, photosynthesis and transpiration in *Psophocarpus tetragonolobus* (L.) DC cultivar "UPS 99". - Photosynthesis Res. 2: 259-268, 1981.

*8910 - LEOPOLD, A.C.: Temperature effects on soybean imbibition and leakage. - Plant Physiol. 65: 1096-1098, 1980.

8911 - LEOPOLD, A.C., MUSGRAVE, M.E., WILLIAMS, K.M.: Solute leakage resulting from leaf desiccation. - Plant Physiol. 68: 1222-1225, 1981.

*8912 - LEUNING, R., NEUMANN, H.H., THURTELL, G.W.: Ozone uptake by corn (*Zea mays* L.): a general approach. - Agr. Meteorol. 20: 115-135, 1979.

8913 - LEVERENZ, J.W.: Photosynthesis and transpiration in large forest-grown Douglas-fir: diurnal variation. - Can. J. Bot. 59: 349-356, 1981.

*8914 - LEVERENZ, J.W., JARVIS, P.G.: Photosynthesis in Sitka spruce (*Picea sitchensis* (Bong.) Carr.). X. Acclimation to quantum flux density within and between trees. - J. appl. Ecol. 17: 697-708, 1980.

8915 - LICHTENTHALER, H.K., BUSCHMANN, C., DÖLL, M., FIETZ, H.-J., BACH, T., KOZEL, U., MEIER, D., RAHMSDORF, U.: Photosynthetic activity, chloroplast ultrastructure, and leaf characteristics of high-light and low-light plants and of sun and shade leaves. - Photosynthesis Res. 2: 115-141, 1981.

8916 - **LIE, T.A.:** Environmental physiology of the legume-*Rhizobium* symbiosis. -
In: BROUGHTON, W.J. (ed.): Nitrogen Fixation. Volume I. Ecology. Pp. 104-134.
Clarendon Press, Oxford 1981.

8917 - **LIEFFERS, V.J., SHAY, J.M.:** The effects of water level on the growth and
reproduction of *Scirpus maritimus* var. *paludosus*. - Can. J. Bot. 59: 118-121,
1981.

8918 - **LILLEY, R.McC., LARKUM, A.W.D.:** Isolation of funcionally intact rhodoplasts
from *Griffithsia monilis* (Ceramiaceae, Rhodophyta). - Plant Physiol. 67:
5-8, 1981.

8919 - **LINDER, S.:** Photosynthesis and respiration in a young stand of Scots pine. -
In: Proc. XVII IUFRO World Congress. Division 2. Pp. 97-108. International
Union Forest Research Organisation, Japan 1981.

8920 - **LINDER, S. (ed.):** Understanding and Predicting Tree Growth. (Studia forest.
Suecica 160 - Models of Tree Growth.) - Swed. Univ. Agr. Sci., Uppsala 1981.

*8921 - **LINDER, S., TROENG, E.:** Photosynthesis and transpiration of 20-year-old Scots
pine. - Ecol. Bull. (Stockholm) 32 (PERSSON, T. (ed.): Structure and Function
of Northern Coniferous Forests - An Ecosystem Study.): 165-181, 1980.

*8922 - **LITTLER, M.M., ARNOLD, K.E.:** Sources of variability in macroalgal primary
productivity: sampling and interpretative problems. - Aquat. Bot. 8: 141-156,
1980.

8923 - **LITTLETON, E.J., DENNETT, M.D., ELSTON, J., MONTEITH, J.L.:** The growth and
development of cowpeas (*Vigna unguiculata*) under tropical field conditions.
3. Photosynthesis of leaves and pods. - J. agr. Sci. 97: 539-550, 1981.

*8924 - **LOACH, K., GAY, A.P.:** The light requirement for propagating hardy ornamental
species from leafy cuttings. - Scientia Hort. 10: 217-230, 1979.

*8925 - **LOHAMMAR, T., LARSSON, S., LINDER, S., FALK, S.O.:** Fast-simulation models
of gaseous exchange in Scots pine. - Ecol. Bull. (Stockholm) 32 (PERSSON, T.
(ed.): Structure and Function of Northern Coniferous Forests - An Ecosystem
Study.): 505-523, 1980.

8926 - **LONGSTRETH, D.J., HARTSOCK, T.L., NOBEL, P.S.:** Light effects on leaf development and photosynthetic capacity of *Hydrocotyle bonariensis* Lam. - Photosynthesis Res. 2: 95-104, 1981.

8927 - **LOON, C.D.,van:** The effect of water stress on potato growth, development, and
yield. - Amer. Potato J. 58: 51-69, 1981.

8928 - **LÖSCH, R., KAPPEN, L.:** The cold resistance of Macaronesian Sempervivoideae. -
Oecologia 50: 98-102, 1981.

8929 - **LÖSCH, R., TENHUNEN, J.D.:** Stomatal responses to humidity - phenomenon and
mechanism. - In: JARVIS, P.G., MANSFIELD, T.A. (ed.): Stomatal Physiology.
Pp. 137-161. Cambridge University Press, Cambridge - London - New York -
New Rochelle - Melbourne - Sydney 1981.

*8930 - **L'ROY, A., HENDRIX, D.L.:** Effect of salinity upon cell membrane potential in
the marine halophyte, *Salicornia bigelovii* Torr. - Plant Physiol. 65: 544-
549, 1980.

8931 - **LUARD, E.J., GRIFFIN, D.M.:** Effect of water potential on fungal growth and
turgor. - Trans. Brit. mycol. Soc. 76: 33-40, 1981.

8932 - **LUCAS, W.J., ALEXANDER, J.M.:** Influence of turgor pressure manipulation on
plasmalemma transport of HCO_3^- and OH^- in *Chara corallina*. - Plant Physiol.
68: 553-559, 1981.

8933 - LUCHKOŬ, A.I., DZYARUGINA, T.F.: Anatamichnaya budova lista seyantsaŭ drèva-
vykh raslin u zalezhnastsi ad vodazabyaspechanastsi. [Anatomical structure of
a leaf of arboreal seedlings depending on the water supply.] - Vestsi Akad.
Navuk Belaruss. SSR, Ser. biyal Navuk 1981(3): 14-17,123, 1981. [In Belorus,
ab: E, R.]

*8934 - LUKINA, L.F.: Vodnyĭ rezhim nekotorykh vidov vysshikh vodnykh rasteniĭ.
[Water regime of some higher aquatic plant species.] - Gidrobiol. Zh. 16(6):
89-90, 1980. [In R.]

8935 - LUMMITSCH, M., LAUBIG, U.: Biophysikalische und numerische Untersuchungen zum
Einfluss exogener Faktoren auf Teilprozesse der Stoffproduktion bei Weizen in
der Kornfüllungsperiode. - In: UNGER, K., STÖCKER, G. (ed.): Biophysikalische
Ökologie und Ökosystemforschung. Pp. 81-92. Akademie-Verlag, Berlin 1981.

*8936 - LUND, V., GOKSØYR, J.: Effects of water fluctuations on microbial mass and
activity in soil. - Microbiol. Ecol. 6: 115-124, 1980.

8937 - LUSH, W.M., GROVES, R.H., KAYE, P.E.: Presowing hydration-dehydration treatments in relation to seed germination and early seedling growth of wheat and
ryegrass. - Aust. J. Plant Physiol. 8: 409-425, 1981.

8938 - MAAREL, E.,van der: Fluctuations in a coastal dune grassland due to fluctuations in rainfall: Experimental evidence. - Vegetatio 47: 259-265, 1981.

8939 - MacMAHON, J.A., SCHIMPF, D.J.: Water as a factor in the biology of North
American desert plants. - In: EVANS, D.D., THAMES, J.L. (ed.): Water in Desert
Ecosystems. Pp. 114-171. Dowden, Hutchinson & Ross, Inc., Stroudsburg 1981.

*8940 - MacROBBIE, E.A.C.: Stomatal ionic relations. - In: SPANSWICK, R.M., LUCAS,
W.J., DAINTY, J. (ed.): Plant Membrane Transport: Current Conceptual Issues.
Pp. 97-107. Elsevier / North-Holland Biomedical Press, Amsterdam - New York
- Oxford 1980.

8941 - MacROBBIE, E.A.C.: Ion fluxes in 'isolated' guard cells of Commelina communis
L. - J. exp. Bot. 32: 545-562, 1981.

8942 - MacROBBIE, E.A.C.: Effects of ABA in 'isolated' guard cells of Commelina
communis L. - J. exp. Bot. 32: 563-572, 1981.

8943 - MacROBBIE, E.A.C.: Ionic relations of stomatal guard cells. - In: JARVIS,
P.G., MANSFIELD, T.A. (ed.): Stomatal Physiology. Pp. 51-70. Cambridge
University Press, Cambridge - London - New York - New Rochelle - Melbourne -
Sydney 1981.

8944 - MAERTENS, C., BLANCHET, R.: Influence des caractères hydriques du milieu
racinaire et aérien sur le potentiel de l'eau dans les feuilles de quelques
types variétaux de soja et confrontation à leur comportement agronomique. -
Agronomie 1: 199-206, 1981.

8945 - MAHLER, R.L., WOLLUM, A.G.,II.: The influence of irrigation and Rhizobium
japonicum strains on yields of soybeans grown in a Lakeland sand. - Agron. J.
73: 647-651, 1981.

8946 - MAHLER, R.L., WOLLUM, A.G.,II.: The influence of soil water potential and
soil texture on the survival of Rhizobium Japonicum and Rhizobium Legumino-
sarum isolates in the soil. - Soil Sci. Soc. Amer. J. 45: 761-766, 1981.

8947 - MAIER-MAERCKER, U.: "Peristomatal transpiration" and stomatal movement:
A controversial view. V. Rubidium-86 in the epidermal transpiration stream. -
Z. Pflanzenphysiol. 101: 447-460, 1981.

8948 - **MAIER-MAERCKER, U.**: "Peristomatal transpiration" and stomatal movement: A controversial view. VII. Correlation of stomatal aperture with evaporative demand and water uptake through the roots. - Z. Pflanzenphysiol. 102: 397-413, 1981.

8949 - **MAIER-MAERCKER, U.**: "Peristomatal transpiration" and stomatal movement: A controversial view. VIII. Stomatal control by conditions of water supply and peristomatal transpiration. - Z. Pflanzenphysiol. 103: 15-25, 1981.

8950 - **MALLOCH, K.R., FENTON, R.**: Reversible effects of farnesol on *Commelina communis*. - New Phytol. 88: 249-254, 1981.

8951 - **MANSFIELD, T.A., DAVIES, W.J.**: Stomata and stomatal mechanisms. - In: PALEG, L.G., ASPINALL, D. (ed.): The Physiology and Biochemistry of Drought Resistance in Plants. Pp. 315-346. Academic Press, Sydney - New York - London - Toronto - San Francisco 1981.

8952 - **MANSFIELD, T.A., TRAVIS, A.J., JARVIS, R.G.**: Responses to light and carbon dioxide. - In: JARVIS, P.G., MANSFIELD, T.A. (ed.): Stomatal Physiology. Pp. 119-135. Cambridge University Press, Cambridge - London - New York - New Rochelle - Melbourne - Sydney 1981.

8953 - **MANSFIELD, T.A., WILSON, J.A.**: Regulation of gas exchange in water-stressed plants. - In: JOHNSON, C.B. (ed.): Physiological Processes Limiting Plant Productivity. Pp. 237-251. Butterworths, London - Boston - Sydney - Wellington - Durban - Toronto 1981.

8954 - **MANSOUR, K.S., HALLET, J.N.**: Effect of desiccation on DNA synthesis and the cell cycle of the moss *Polytrichum formosum*. - New Phytol. 87: 315-324, 1981.

*8955 - **MANTELL, A., GOLDSCHMIDT, E.E., MONSELISE, S.P.**: Turnover of tritiated water in Calamondin plants: fruit-leaf competition. - Plant Physiol. 65 (Suppl.): 9, 1980.

8956 - **MARKHART, A.H.,III, SIONIT, N., SIEDOW, J.N.**: Cell wall water dilution: an explanation of apparent negative turgor potentials. - Can. J. Bot. 59: 1722-1725, 1981.

8957 - **MARSCHNER, H., KYLIN, A., KUIPER, P.J.C.**: Differences in salt tolerance of three sugar beet genotypes. - Physiol. Plant. 51: 234-238, 1981.

*8958 - **MARSHO, T.V., SOKOLOVE, P.M., MACKAY, A.B.**: Regulation of photosynthetic electron transport in intact spinach chloroplasts. I. Influence of exogenous salts on oxaloacetate reduction. - Plant Physiol. 65: 703-706, 1980.

8959 - **MARTIGNONE, R.A., NAKAYAMA, F.**: Efecto de tres regímenes hídricos sobre el crecimiento y componentes del rendimiento de plantas de soja. [Effects of three water regimes on the growth and yield components of soybean plants.] - Fyton 41: 103-113, 1981. [In Span, ab: E.]

8960 - **MARTIN, B., ORT, D.R., BOYER, J.S.**: Impairment of photosynthesis by chilling-temperatures in tomato. - Plant Physiol. 68: 329-334, 1981.

8961 - **MARTIN, C.E., SIEDOW, J.N.**: Crassulacean Acid Metabolism in the epiphyte *Tillandsia usneoides* L. (Spanish moss). Responses of CO_2 exchange to controlled environmental conditions. - Plant Physiol. 68: 335-339, 1981.

*8962 - **MARTIN, G.C., URIU, K., NISHIJIMA, C.**: The effect of drastic reduction of water input on mature walnut trees. - HortScience 15: 157-158, 1980.

8963 - MARTIN, R.J.: Yield - tenderometr relationships in vining peas. - N.Zeal. J. exp. Agr. 9: 387-391, 1981.

8964 - MARTIN, R.J., TABLEY, F.J.: Effects of irrigation, time of sowing, and cultivar on yield of vining peas. - N. Zeal. J. exp. Agr. 9: 291-297, 1981.

*8965 - MARUYAMA, K.: [The Shimadzu climatized chamber for measuring photosynthesis, respiration and transpiration in pot grown Japanese beech seedlings.] - Bull. Niigata Univ. Forest 1980 (13): 1-22, 1980. [In Jap, ab: E.]

8966 - MASON, W.K., SMITH, R.C.G.: Irrigation for crops in a sub-humid environment. III. An irrigation scheduling model for predicting soybean water use and crop yield. - Irrig. Sci. 2: 89-101, 1981.

8967 - MASON, W.K., SMITH, R.C.G.: Irrigation for crops in a sub-humid environment. IV. Analysis of current irrigation practice of soybeans and the potential for improved efficiency. - Irrig. Sci. 2: 103-111, 1981.

8968 - MATHENY, T.A., HUNT, P.G.: Effects of irrigation and sulphur application on soybeans grown on a Norfolk loamy sand. - Commun. Soil Sci. Plant Anal. 12: 147-159, 1981.

*8969 - MATORIN, D.N., ORTOIDZE, T.V., NIKOLAEV, G.M., VENEDIKTOV, P.S.: Sostoyanie vody i funkstionirovanie fotosinteticheskikh reaktsiĭ v khloroplastakh pri razlichnoĭ vlazhnosti. [The state of water and the functioning of photosynthetic reactions in chloroplasts at different humidity.] - Nauch. Dokl. vyssh. Shkoly, biol. Nauki 1980 (10): 17-24, 1980. [In R.]

8970 - MATSUDA, K., RIAZI, A.: Stress-induced osmotic adjustment in growing regions of barley leaves. - Plant Physiol. 68: 571-576, 1981.

*8971 - MATSUI, T., EGUCHI, H., MORI, K., TATEISHI, J., NONAMI, H.: Humidity distributions in plant population under different air currents. - Environ. Control Biol. 18(2): 29-37, 1980.

8972 - MAURYA, P.R., LAL, R.: Effects of different mulch materials on soil properties and on the root growth and yield of maize (Zea mays) and cowpea (Vigna unguiculata). - Field Crops Res. 4: 33-45, 1981,

8973 - MAYORAL, M.L., ATSMON, D., SHIMSHI, D., GROMET-ELHANAN, Z.: Effect of water stress on enzyme activities in wheat and related wild species: carboxylase activity, electron transport and photophosphorylation in isolated chloroplasts. - Aust. J. Plant Physiol. 8: 385-393, 1981.

*8974 - MAZZA, G.: Water vapor equilibrium relationships of potato slices. - Amer. Potato J. 57: 91-100, 1980.

8975 - McAULIFFE, D., APPLEBY, A.P.: Effect of a pre-irrigation period on the activity of ethofumesate applied to dry soil. - Weed Sci. 29: 712-717, 1981.

8976 - McCLENDON, J.H.: The balance of forces generated by the water potential in the cell-wall-matrix - a model. - Amer. J. Bot. 68: 1263-1268, 1981.

8977 - McCLENDON, J.H.: The osmotic pressure of concentrated solutions of polyethylene glycol 6000, and its variation with temperature. - J. exp. Bot. 32: 861-866, 1981.

8978 - McCOWN, R.L., WALL, B.H.: The influence of weather on the quality of tropical legume pasture during the dry season in Northern Australia. II. Moulding of standing hay in relation to rain and dew. - Aust. J. agr. Res. 32: 589-598, 1981.

8979 - McCOWN, R.L., WALL, B.H., HARRISON, P.G.: The influence of weather on the quality of tropical legume pasture during the dry season in Northern Australia. I. Trends in sward structure and moulding of standing hay at three locations. - Aust. J. agr. Res. 32: 575-587, 1981.

8980 - McDONALD, A.J.S., JORDAN, J.R., FORD, E.D.: An automated potometer. - J. exp. Bot. 32: 581-589, 1981.

8981 - McGEE, A.B., SCHMIERBACH, M.R., BAZZAZ, F.A.: Photosynthesis and growth in populations of *Populus deltoides* from contrasting habitats. - Amer. Midl. Natur. 105: 305-311, 1981.

8982 - McGRAW, D.C., UNGAR, I.A.: Growth and survival of the halophyte *Salicornia europaea* L. under saline field conditions. - Ohio J. Sci. 81: 109-113, 1981.

8983 - McINTOSH, M.S., MILLER, D.A.: Genetic and soil moisture effects on the branching-root trait in alfalfa. - Crop Sci. 21: 15-18, 1981.

8984 - McINTYRE, G.I.: Apical dominance in the rhizome of *Agropyron repens*: the influence of humidity and light on the regenerative growth of isolated rhizomes. - Can. J. Bot. 59: 549-555, 1981.

*8985 - McMICHAEL, B.L.: Water stress adaptation. - In: HESKETH, J.D., JONES, J.W. (ed.): Predicting Photosynthesis for Ecosystem Models. Volume I. Pp. 183-203. CRC Press, Boca Raton 1980.

8986 - McNEIL, P.H., WALKER, D.A.: The effect of magnesium and other ions on the distribution of ribulose 1,5-bisphosphate carboxylase in chloroplast extracts. - Arch. Biochem. Biophys. 208: 184-188, 1981.

8987 - MEDINA, E., OSMOND, C.B.: Temperature dependence of dark CO_2 fixation and acid accumulation in *Kalanchoë daigremontiana*. - Aust. J. Plant Physiol. 8: 641-649, 1981.

*8988 - MEETEREN, U., van: Water relations and keeping-quality of cut *Gerbera* flowers. III. Water content, permeability and dry weight of ageing petals. - Scientia Hort. 10: 261-269, 1979.

*8989 - MEETEREN, U., van: Water relations and keeping-quality of cut *Gerbera* flowers. IV. Internal water relations of ageing petal-tissue. - Scientia Hort. 11: 83-93, 1979.

8990 - MEIDNER, H.: Measurements of stomatal aperture and responses to stimuli. - In: JARVIS, P.G., MANSFIELD, T.A. (ed.): Stomatal Physiology. Pp. 25-49. Cambridge University Press, Cambridge - London - New York - New Rochelle - Melbourne - Sydney 1981.

8991 - MEIDNER, H.: What next? - In: JARVIS, P.G., MANSFIELD, T.A. (ed.): Stomatal Physiology. Pp. 281-286. Cambridge University Press, Cambridge - London - New York - New Rochelle - Melbourne - Sydney 1981.

*8992 - MEIRI, A.: Transpiration and water status parameters of sunflower leaves absorbing chloride or sulfate solutions. - Plant Physiol. 65 (Suppl.): 82, 1980.

8993 - MENGEL, D.B., WILSON, F.E.: Water management and nitrogen fertilization of ratoon crop rice. - Agron. J. 73: 1008-1010, 1981.

8994 - MÉRIAUX, S., ROLLIN, H., RUTTEN, P.: Effets de la sécheresse sur la vigne (*Vitis vinifera* L.), II. - Etudes sur "Grenache". - Agronomie 1: 375-382, 1981.

8995 - MÉRIDA, T., SCHÖNHERR, J., SCHMIDT, H.W.: Fine structure of plant cuticles in relation to water permeability: The fine structure of the cuticle of *Clivia miniata* Reg. leaves. - Planta 152: 259-267, 1981.

*8996 - METZ, S.G., SCHAREN, A.L.: Potential for the development of *Pyrenophora graminea* on barley in a semi-arid environment. - Plant Dis. Rep. 63: 671-675, 1979.

*8997 - METZLER, J.T., PELL, E.J.: The impact of peroxyacetyl nitrate on conductance of bean leaves and on associated cellular and foliar symptom expression. - Phytopathology 70: 934-938, 1980.

8998 - MEYER, R.F., BOYER, J.S.: Osmoregulation, solute distribution, and growth in soybean seedlings having low water potentials. - Planta 151: 482-489, 1981.

8999 - MEYER, W.S., GREEN, G.C.: Comparison of stomatal action of orange, soybean and wheat under field conditions. - Aust. J. Plant Physiol. 8: 65-76, 1981.

9000 - MEYER, W.S., GREEN, G.C.: Field test of an irrigation scheduling computer model. - Water S.A. 7: 43-48, 1981.

9001 - MEYER, W.S., GREEN, G.C.: Plant indicators of wheat and soybean crop water stress. - Irrig. Sci. 2: 167-176, 1981.

*9002 - MICHELENA, V.A., BOYER, J.S.: Growth and osmotic adjustment in maize leaves having low water potentials. - Plant Physiol. 65 (Suppl.): 8, 1980.

9003 - MIDDLETON, J.E., PROEBSTING, E.L., ROBERTS, S.: A comparison of trickle and sprinkler irrigation for apple orchards. - Washington State Univ., Coll. Agr. Res. Cent. Bull. 0895: 1-5, 1981.

*9004 - MIGLIACCIO, F.: Persistence of modifications induced by osmo-saline shocks on chloride transport and some related metabolic processes in roots of maize plants. - J. exp. Bot. 31: 1555-1564, 1980.

9005 - MILBORROW, B.V.: Abscisic acid and other hormones. - In: PALEG, L.G., ASPINALL, D. (ed.): The Physiology and Biochemistry of Drought Resistance in Plants. Pp. 347-388. Academic Press, Sydney - New York - London - Toronto - San Francisco 1981.

*9006 - MILES, L.J., PARKER, G.R.: Effects of cadmium and one-time drought stress on survival, growth, and yield of native plant species. - J. environ. Qual. 9: 278-283, 1980.

9007 - MILLER, C.A., DAVIS, D.D.: Effect of temperature on stomatal conductance and ozone injury of pinto bean leaves. - Plant Dis. 65: 750-751, 1981.

9008 - MILLER, I.L., WILLIAMS, W.T.: Tolerance of some tropical legumes to six months of simulated waterlogging. - Trop. Grasslands 15: 39-43, 1981.

9009 - MILLER, N.A.: The effect of N-decenylsuccinic acid on the leaf water balance of *Zea mays* L. - Bot. Gaz. 142: 197-199, 1981.

9010 - MILLER, P.C. (ed.): Resource Use by Chaparral and Matorral. A Comparison of Vegetation Function in Two Mediterranean Type Ecosystems. (Ecological Studies Volume 39). - Springer-Verlag, New York - Heidelberg - Berlin 1981,

9011 - MILLER, P.C.: Similarities and limitations of resource utilization in mediterranean type ecosystems. - In: MILLER, P.C. (ed.): Resource Use by Chaparral and Matorral. A comparison of Vegetative Function in Two Mediterranean Type Ecosystems. Pp. 369-407. Springer-Verlag, New York - Heidelberg - Berlin 1981.

9012 - **MILLER, P.C., HAJEK, E.:** Resource availability and environmental characte-
ristics of mediterranean type ecosystems. - In: MILLER, P.C. (ed.): Resource
Use by Chaparral and Matorral. A Comparison of Vegetation Function in Two
Mediterranean Type Ecosystems. Pp. 17-41. Springer-Verlag, New York -
Heidelberg - Berlin 1981.

9013 - **MILLER, P.C., HAJEK, E., POOLE, D.K., ROBERTS, S.W.:** Microclimate and energy
exchange. - In: MILLER, P.C. (ed.): Resource Use by Chaparral and Matorral.
A Comparison of Vegetation Function in Two Mediterranean Type Ecosystems.
Pp. 97-121. Springer-Verlag, New York - Heidelberg - Berlin 1981.

9014 - **MILLER, R.J., ROLSTON, D.E., RAUSCHKOLB, R.S., WOLFE, D.W.:** Labeled nitrogen
uptake by drip-irrigated tomatoes. - Agron. J. 73: 265-270, 1981.

9015 - **MINOGUE, K.P., FRY, W.E.:** Effect of temperature, relative humidity, and
rehydration rate on germination of dried sporangia of *Phytophthora infestans*.
- Phytopathology 71: 1181-1184, 1981.

9016 - **MIRANDA, V., BAKER, N.R., LONG, S.P.:** Anatomical variation along the length
of the *Zea mays* leaf in relation to photosynthesis. - New Phytol. 88: 595-
605, 1981.

*9017 - **MIRCHINK, T.G., SUDNITSYN, I.I., GENDZHIEV, M.G.:** Ustoĭchivost' gribov raznykh mestoobitaniĭ k razlichnoĭ aktivnosti vlagi. [Stability of fungi of
different sites to different moisture activity.] - Pochvovedenie 1978 (6):
55-58, 1978. [In R, ab: E.]

9018 - **MIRHADI, M.J., KOBAYASHI, Y.:** Studies on the productivity of grain sorghum.
IV. Effect of various planting dates on the growth, grain yield and protein
content of irrigated and nonirrigated grain sorghum. - Jap. J. Crop Sci. 50:
115-124, 1981.

9019 - **MIRHADI, M.J., KOBAYASHI, Y.:** Studies on the productivity of grain sorghum.
V. Effect of nitrogen fertilization and water stress on the grain yield,
nitrogen uptake and translocation. - Jap. J. Crop Sci. 50: 131-142, 1981.

9020 - **MISHKIND, M., PALEVITZ, B.A., RAIKHEL, N.V.:** Cell wall architecture: normal
development and environmental modification of guard cells of the *Cyperaceae*
and related species. - Plant Cell Environ. 4: 319-328, 1981.

9021 - **MISLEVY, P., EVERETT, P.H.:** Subtropical grass species response to different
irrigation and harvest regimes. - Agron. J. 73: 601-604, 1981.

9022 - **MISRA, R.D., PANT, P.C.:** Criteria for scheduling the irrigation of wheat. -
Exp. Agr. 17: 157-162, 1981.

*9023 - **MIŠTINOVÁ, A., MIŠTINA, T.:** Variabilita počtu chloroplastov v stomatických
bunkách lucerny (*Medicago sativa* L.). [Variability in chloroplast number
in stomata cells of alfalfa (*Medicago sativa* L.).] - Vedecké Práce výskum.
Ústavu rast. Výroby Piešťanoch 16: 29-38, 1979. [In Slovak, ab: E, R.]

9024 - **MITCHELL, W.H.:** Subsurface irrigation and fertilization of field corn. -
Agron. J. 73: 913-916, 1981.

9025 - **MOELJOPAWIRO, S., IKEHASHI, H.:** Inheritance of salt tolerance in rice. -
Euphytica 30: 291-300, 1981.

9026 - **MOKRONOSOV, A.T.:** Ontogeneticheskiĭ Aspekt Fotosinteza. [Ontogenetic Aspect
of Photosynthesis.] - Nauka, Moskva 1981. [In R.]

9027 - **MOLL, A.:** Untersuchungen zur Modifikation der Ertragsbildung der Kartoffel durch Umwelteinflüsse 1. Mitteilung: Zum Einfluss von Stickstoff und Wasser auf die Ertragsbildung. - Arch. Acker- Pflanzenbau Bodenk. 25: 457-463, 1981.

9028 - **MOLL, A.:** Untersuchungen zur Modifikation der Ertragsbildung der Kartoffel durch Umwelteinflüsse 2. Mitteilung: Zum Jahreseinfluss auf die Ertrags- bildung. - Arch. Acker- Pflanzenbau Bodenk. 25: 465-475, 1981.

9029 - **MØLLER, E., NIELSEN, C., RASMUSSEN, K.J.:** Genvaekst efter fortørring af graesmarksafgrøder. I. Daekningseffekt. [Regrowth after pre-wilting of grass- land crops. I. The effect of covering.] - Tidsskr. Planteavl 83: 497-504, 1979. [In Dan, ab: E.]

9030 - **MOLZ, F.J.:** Models of water transport in the soil-plant system: a review. - Water Resour. Res. 17: 1245-1260, 1981.

*9031 - **MONSON, R.K., SMITH, S.D.:** Seasonal water relations components of native desert plants. - Plant Physiol. 65 (Suppl.): 48, 1980.

*9032 - **MONSON, R.K., WILLIAMS, G.J.,III:** Photosynthetic acclimation to temperature in *Carex stenophylla*. - Plant Physiol. 65(Suppl.): 48, 1980.

*9033 - **MONTEITH, J.L.:** Coupling of plants to the atmosphere. - In: GRACE, J., FORD, E.D., JARVIS, P.G. (ed.): Plants and their Atmospheric Environment. Pp. 1-29. Blackwell Scientific Publications, Oxford - London - Edinburgh - Boston - Melbourne 1980.

*9034 - **MONTEITH, J.L., CAMPBELL, G.S.:** Diffusion of water vapour through integuments - Potential confusion. - J. therm. Biol. 5: 7-10, 1980.

*9035 - **MONTENEGRO, G., RIVEROS, F., ALCALDE, C.:** Morphological structure and water balance of four chilean shrub species. - Flora 170: 554-564, 1980.

9036 - **MONTENY, B.A., HUMBERT, J., LHOMME, J.P., KALMS, J.M.:** Le rayonnement net et l'estimation de l'evapotranspiration en Côte d'Ivoire. - Agr. Meteorol. 23: 45-59, 1981.

9037 - **MOORBY, J.:** Transport Systems in Plants. - Longman, London - New York 1981.

9038 - **MOORE, P.:** The varied ways plants tap the Sun. - New Scientist 89: 394-397, 1981.

9039 - **MOORE, T.C.:** Research Experiences in Plant Physiology. 2nd Edition. - Sprin- ger-Verlag, New York - Heidelberg - Berlin 1981.

9040 - **MORANDI, E.N., REGGIARDO, L.M., NAKAYAMA, F.:** Efectos del cloruro de (2-clo- roetil)trimetilamonio (CCC) sobre el crecimiento vegetativo y reproductivo y el consumo de agua en soja (*Glycine max* (L.) Merr.) cultivada con alta disponibilidad hídrica. [Effects of 2-chloroethyl trimethylammonium chloride (CCC) on vegetative and reproductive growth and water consumption in the soybean (*Glycine max* (L.) Merr.) cultivated at high humidity.] - Fyton 41: 115-128, 1981. [In Span, ab: E.]

9041 - **MORESHET, S.:** Physiological activity, in a semiarid environment, of *Eucalyp- tus camaldulensis* Dehn. from two provenances. - Aust. J. Bot. 29: 97-110, 1981.

*9042 - **MORESHET, S., COHEN, Y., FUCHS, M.:** Effect of increasing foliage reflectance on yield, growth, and physiological behavior of a dryland cotton crop. - Crop Sci. 19: 863-868, 1979.

*9043 - MORIZET, J.: Étude des transferts de l'eau chez le Tournesol (*Helianthus annuus*). Analyse de l'interaction entre alimentation azotée et aération du milieu racinaire. - C.R. Acad. Sci. Paris, Sér. D 290: 1229-1232, 1980.

*9044 - MORK, H.M., WALLACE, A., ROMNEY, E.M.: Effect of certain plant parameters on photosynthesis, transpiration, and efficiency of water use. - Great Basin Naturalist Memoirs 1980 (4 - Soil-Plant-Animal Relationships Bearing on Revegetation and Land Reclamation in Nevada Deserts): 117-120, 1980.

9045 - MORRIS, J.D.: Factors affecting the salt tolerance of eucalypts. - In: OLD, K.M., KILE, G.A., OHMART, C.P. (ed.): Eucalypt Dieback in Forests and Woodlands. Pp. 190-204. CSIRO, South Carlton 1981.

9046 - MORRISON, J.E.,Jr., BARFIELD, B.J.: Temperature predictions for plant and controlled-resistance artificial leaves. - Trans. ASAE 24: 1204-1210, 1981.

9047 - MORRISON, J.E.,Jr., BARFIELD, B.J.: Transpiring artificial leaves. - Agr. Meteorol. 24: 227-236, 1981.

*9048 - MOTHA, R.P., VERMA, S.B., ROSENBERG, N.J.: Flux-anlage distribution of momentum and sensible heat over a vegetated surface. - Agr. Meteorol. 22: 121-127, 1980.

9049 - MOURAVIEFF, I.: Réduction *in vivo* du Nitro bleu de tétrazolium par les cellules des épidermes détachés. Effets d'une longue privation à l'obscurité. - Ann. Sci. natur. bot. (Paris) 13: 93-97, 1980-1981.

9050 - MOUSSEAU, M.: Effect of a change in photoperiod on subsequent CO_2 exchange. - Photosynthesis Res. 2: 85-94, 1981.

9051 - MOUTONNET, P., WERY, J., BOSSY, A.: Etude comparative de quatre dispositifs d'irrigation automatique: application au maïs irrigué en localisé. - Agr. Meteorol. 24: 275-289, 1981.

*9052 - MOZHAEVA, L.V., PIL'SHCHIKOVA, N.V.: O dvizhushcheĭ sile placha rasteniĭ. [The moving force of plant exudation.] - Fiziol. Rast. 26: 994-1000, 1979. [In R, ab: E.]

*9053 - MROZEK, E.,Jr.: Effect of mercury and cadmium on germination of *Spartina alterniflora* Loisel. seeds at various salinities. - Environ. exp. Bot. 20: 367-377, 1980.

9054 - MTUI, T.A., KANEMASU, E.T., WASSOM, C.: Canopy temperatures, water use, and water use efficiency of corn genotypes. - Agron. J. 73: 639-643, 1981.

9055 - MUCHOW, R.C., WOOD, I.M.: Pattern of infiltration with furrow irrigation and evapotranspiration of kenaf (*Hibiscus cannabinus*) grown on Cununurra clay in the Ord Irrigation Area. - Aust. J. exp. Agr. anim. Husb. 21: 101-108, 1981.

9056 - MUIRHEAD, W.A., WHITE, R.J.G.: The influence of soil water potential on the flowering pattern, pod set and yield of snap beans (*Phaseolus vulgaris* L.). - Irrig. Sci. 3: 45-56, 1981.

9057 - MUKHERJEE, S.P., CHOUDHURI, M.A.: Effect of water stress on some oxidative enzymes and senescence in *Vigna* seedlings. - Physiol. Plant. 52: 37-42, 1981.

9058 - MUĽÁR, J., DERCO, M.: Účinok vápnenia ťažkej odvodnenej pôdy na úrodu cukrovej repy pri použití závlahy. [The effect of the liming of heavy-textured drained soil on the yield of irrigated sugar-beet.] - Rost. Výroba (Praha) 27: 1103-1112, 1981. [In Slov, ab: E, R, G.]

*9059 - MULEV, M.: Vlijanie na stepenot na degradatsijata na zaednitsata *Coccifero-*
        -Carpinetum orientalis (Oberd. 1948) em Ht. 1954, VRZ osmotskiot pritisok
        na *Quercus coccifera* L. i *Phillyrea media* L. [Influence of degree of degra-
        dation of ass. *Coccifero-Carpinetum orientalis* (Oberd. 1948) em Ht. 1954,
        upon osmotic pressure of *Quercus coccifera* L. and *Phillyrea media*.L.]   -
        God. Zb. biol. Fak. Univ. "Kiril Metodij"(Skopje) 32: 151-161, 1979.
        [In Macedon, ab: E.]

 9060 - MÜLLER, L., STARNECKER, G., WINKLER, S.: Zur Ökologie epiphytischer Farne in
        Südbrasilien I.Saugschuppen. - Flora 171: 55-63, 1981.

*9061 - MÜLLER, W.: Zum "Obstgarten-Klima" (Strahlung und Temperatur) in Österreich.
        - Pflanzenschutz-Berichte 45: 97-127, 1979.

 9062 - MULLINS, J.A.: Estimation of the plant available water capacity of a soil
        profile. - Aust. J. Soil Res. 19: 197-207, 1981.

 9063 - MUNNS, R., WEIR, R.: Contribution of sugars to osmotic adjustment in elonga-
        ting and expanded zones of wheat leaves during moderate water deficits at
        two light levels. - Aust. J. Plant Physiol. 8: 93-105, 1981.

 9064 - MURPHY, C.E.,Jr., SCHUBERT, J.F., DEXTER, A.H.: The energy and mass exchange
        characteristics of a loblolly pine plantation. - J. appl. Ecol. 18: 271-281,
        1981.

*9065 - NABORS, M.W., GIBBS, S.E., BERNSTEIN, C.S., MEIS, M.E.: NaCl-tolerant tobacco
        plants from cultured cells. - Z. Pflanzenphysiol. 97: 13-17, 1980.

 9066 - NAGARAJAH, S.: The effect of nitrogen on plant water relations in tea (*Ca-
        mellia sinensis*). - Physiol. Plant. 51: 304-308, 1981.

 9067 - NAGARAJAN, S., PANDA, B.C.: Dynamics of protein and bound water formation
        in leaf tissues of wheat. - Ind. J. exp. Biol. 19: 773-775, 1981.

 9068 - NAGY, J., ZEKE, É.: A kukoricaszemek vízleadásának vizsgálata I. A műtrá-
        gyázás hatása a szemnedvességre. [Study of water release by maize grains. I.
        Effect of fertilization on grain moisture.] - Növénytermelés 30: 529-537,
        1981. [In Hung, ab: E.]

*9069 - NAKAGAWA, S., KOMATSU, H., YUDA, E.: A study of micro-morphology of grape
        berry surface during their development with special reference to stoma. -
        J. Jap. Soc. hort. Sci. 49: 1-7, 1980.

*9070 - NARASAGOUDAR, N.A., CHAVAN, P.D., KARADGE, B.A.: Germination of *Sorghum
        vulgare* Pers. under saline conditions. - Geobios 6: 327-328, 1979.

 9071 - NASS, H.G., STERLING, J.D.E.: Comparison of tests characterizing varieties
        of barley and wheat for moisture stress resistance. - Can. J. Plant Sci.
        61: 283-292, 1981.

 9072 - NAVARA, J.: Transpirationsintensität einiger Gehölze unter Immissionsbedin-
        gungen. - Biológia (Bratislava) 36: 261-267, 1981.

 9073 - NELL, T.A., ALLEN, J.J., JOINER, J.N., ALBRIGO, L.G.: Light, fertilizer, and
        water level effects on growth, yield, nutrient composition, and light compen-
        sation point of *Chrysanthemum*. - HortScience 16: 222-224, 1981.

 9074 - NERKAR, Y.S., WILSON, D., LAWES, D.A.: Genetic variation in stomatal characte-
        ristics and behaviour, water use and growth of five *Vicia faba* L. genotypes
        under contrasting soil moisture regimes. - Euphytica 30: 335-345, 1981.

9075 - **NIEUWENHUIS, G.J.A.:** Application of HCMM satellite and airplane reflection and heat maps in agrohydrology. - Tech. Bull. (Wageningen) 122: 71-86, 1981.

*9076 - **NIKOLAEVA, M.G., POLYAKOVA, E.I., RAZUMOVA, M.V., ASKOCHENSKAYA, N.A.:** O mekhanizme tormozheniya prorastaniya v semenakh zheltoĭ akatsii. [On the mechanism of inhibition of seed germination of siberian pea-tree (*Caragana arborescens*).] - Fiziol. Rast. 25: 1251-1260, 1978. [In R, ab: E.]

9077 - **NILSEN, E.T., MULLER, W.H.:** The influence of low plant water potential on the growth and nitrogen metabolism of the native California shrub *Lotus scoparius* (Nutt. in T & G) Ottley. - Amer. J. Bot. 68: 402-407, 1981.

9078 - **NILSEN, E.T., SCHLESINGER, W.H.:** Phenology, productivity, and nutrient accumulation in the post-fire chaparral shrub *Lotus scoparius*. - Oecologia 50: 217-224, 1981.

*9079 - **NIRALE, A.S., GAUR, B.K.:** Differential effects of X-irradiation on the growth and water uptake in bean seedlings. - Ind. J. Plant Physiol. 23: 244-247, 1980.

9080 - **NISHIZAWA, N., MORI, S.:** A vesicle containing fibrous materials appearing in water-stressed corn root cells. - Plant Cell Physiol. 22: 933-938, 1981.

*9081 - **NITO, N., KURAISHI, S., SUMINO, T.:** Daily changes in the highest leaf surface temperature of plants growing at Heiwa Avenue, Hiroshima. - Environ. Control Biol. 17: 59-66, 1979.

9082 - **NOBEL, P.S.:** Spacing and transpiration of various sized clumps of a desert grass, *Hilaria rigida* - J. Ecol. 69: 735-742, 1981.

9083 - **NOBEL, P.S.:** Wind as an ecological factor. - In: LANGE, O.L., NOBEL, P.S., OSMOND, C.B., ZIEGLER, H. (ed.): Physiological Plant Ecology I. Responses to the Physical Environment. Pp. 475-500. Springer-Verlag, Berlin - Heidelberg - New York 1981.

9084 - **NOBEL, P.S., HARTSOCK, T.L.:** Development of leaf thickness for *Plectranthus parviflorus* - Influence of photosynthetically active radiation. - Physiol. Plant. 51: 163-166, 1981.

9085 - **NOBEL, P.S., HARTSOCK, T.L.:** Shifts in the optimal temperature for nocturnal CO_2 uptake caused by changes in growth temperature for cacti and agaves. - Physiol. Plant. 53: 523-527, 1981.

*9086 - **NOODĚN, L.D.:** Senescence in the whole plant. - In: THIMANN, K.V. (ed.): Senescence in Plants. Pp. 219-258. CRC Press, Boca Raton 1980.

9087 - **NORBY, R.J., KOZLOWSKI, T.T.:** Relative sensitivity of three species of woody plants to SO_2 at high or low exposure temperature. - Oecologia 51: 33-36, 1981.

9088 - **NORDEN, A.J.:** Effect of preparation and storage environment on lifespan of shelled peanut seed. - Crop Sci. 21: 263-266, 1981.

*9089 - **NORMAN, J.M., HESKETH, J.D.:** Micrometeorological methods for predicting environmental effects on photosynthesis. - In: HESKETH, J.D., JONES, J.W. (ed.): Predicting Photosynthesis for Ecosystem Models. Vol. I. Pp. 9-35. CRC Press, Boca Raton 1980.

9090 - **NOSE, A., MIYAZATO, K., MURAYAMA, S.:** Studies on matter production in pineapple plant. II. Effects of soil moisture on the gas exchange of pineapple plants. - Jap. J. Crop Sci. 50: 525-535, 1981.

9091 - **NOVOA, R., LOOMIS, R.S.:** Nitrogen and plant production. - Plant Soil 58: 177-204, 1981. Also in: MONTEITH, J., WEBB, C. (ed.): Soil Water and Nitrogen in Mediterranean-Type Environments. Development in Plant and Soil Sciences. Volume 1. Martinus Nijhoff / Dr. W. Junk Publishers, The Hague - Boston - London 1981.

9092 - **NOWAKOWSKI, W.:** Action de l'acide indolyl-3-acétique sur la déshydratation des limbes chez quatre espèces de blé avec différent niveau de ploïdie. - Biol. Plant. 23: 302-305, 1981.

9093 - **NUKAYA, A., MASUI, M., ISHIDA, A.:** Relationships between salt tolerance of green soabeans and calcium sulfate applications in sand culture. - J. Jap. Soc. hort. Sci. 50: 326-331, 1981.

9094 - **OBERBAUER, S.F., BILLINGS, W.D.:** Drought tolerance and water use by plants along an alpine topographic gradient. - Oecologia 50: 325-331, 1981.

9095 - **OBŁOJ, H., KACPERSKA, A.:** Desiccation tolerance changes in winter rape leaves grown under different environmental conditions. - Biol. Plant. 23: 209-213, 1981.

*9096 - **ODVODY, G.N., DUNKLE, L.D.:** Charcoal stalk rot of sorghum: Effect of environment on host-parasite relations. - Phytopathology 69: 250-254, 1979.

*9097 - **ODZHATING, I.:** Strukturnaya adaptatsiya k vodnomu faktoru list'ev osnovnykh mangrovoobrazuyushchikh drevesnykh rasteniĭ Nigerii. [Structural adaptation of the major mangrove woody plants of Nigeria to the water factor of leaves.] - Bot. Zh. 65: 1784-1788, 1980. [In R.]

9098 - **OEBKER, N.F.:** Vegetable crops in desert areas - problems, practices, and potentials. - In: MANASSAH, J.T., BRISKEY, E.J. (ed.): Advances in Food Producing Systems for Arid and Semiarid Lands, Part B. Pp. 755-771. Academic Press, New York - San Francisco - London 1981.

9099 - **OECHEL, W.C., LAWRENCE, W., MUSTAFA, J., MARTÍNEZ, J.:** Energy and carbon acquisition. - In: MILLER, P.C. (ed.): Resource Use by Chaparral and Matorral. A comparison of Vegetation Function in Two Mediterranean Type Ecosystemes. Pp. 151-183. Springer-Verlag, New York - Heidelberg - Berlin 1981.

9100 - **OEVER, L., van den, BAAS, P., ZANDEE, M.:** Comparative wood anatomy of *Symplocos* and latitude and altitude of provenance. - IAWA Bull. 2: 3-24, 1981.

9101 - **OGAWA, T.:** Blue light response of stomata with starch-containing (*Vicia faba*) and starch-deficient (*Allium cepa*) guard cells under background illumination with red light. - Plant Sci. Lett. 22: 103-108, 1981.

9102 - **OHKI, K.:** Manganese critical levels for soybean growth and physiological processes. - J. Plant Nutr. 3: 271-284, 1981.

*9103 - **OKUSANYA, O.T.:** An experimental investigation into the ecology of some maritime cliff species. - J. Ecol. 67: 579-590, 1979.

*9104 - **OKUSANYA, O.T.:** Quantitative analysis of the effects of photoperiod, temperature, salinity and soil types on the germination and growth of *Corchorus olitorius*. - Oikos 33: 444-450, 1979.

*9105 - **OKUSANYA, O.T.:** The effect of salinity and nutrient level on the growth of *Lavatera arborea*. - Oikos 35: 49-54, 1980.

9106 - **O'LEARY, J.W., KNECHT, G.N.:** Elevated CO_2 concentration increases stomate numbers in *Phaseolus vulgaris* leaves. - Bot. Gaz. 142: 438-441, 1981.

9107 - **O'LEARY, M.H.:** Carbon isotope fractionation in plants. - Phytochemistry 20: 553-567, 1981.

*9108 - **OLIVEIRA, E.C., HSIAO, T.C.**: Osmotic adjustment of cotton to water stress:
Time course and associated changes in growth and assimilation. - Plant Physiol. 65 (Suppl.): 6, 1980.

9109 - **OLSZYK, D.M., TIBBITTS, T.W.**: Stomatal response and leaf injury of *Pisum sativum* L. with SO_2 and O_3 exposures. I. Influence of pollutant level and leaf
maturity. - Plant Physiol. 67: 539-544, 1981.

9110 - **OLSZYK, D.M., TIBBITTS, T.W.**: Stomatal response and leaf injury of *Pisum sativum* L. with SO_2 and O_3 exposures. II. Influence of moisture stress and time
of exposure. - Plant Physiol. 67: 545-549, 1981.

*9111 - **OMASA, K., ABO, F.**: Analysis of air pollutant sorption by plants. (1) Relation
between local SO_2 sorption and acute visible leaf injury. - Res. Rep. nat.
Inst. environ. Studies 11 (Studies on the Effects of Air Pollutants on Plants
and Mechanisms of Phytotoxicity.): 181-193, 1980.

*9112 - **OMASA, K., ABO, F., FUNADA, S., AIGA, I.**: Analysis of air pollutant sorption
by plants. (2) A method for simultaneous measurement of NO_2 and O_3 sorptions
by plants in environmental control chamber. - Res. Rep. nat. Inst. environ.
Studies 11 (Studies on the Effects of Air Pollutants on Plants and Mechanisms
of Phytotoxicity.): 195-211, 1980.

*9113 - **OMASA, K., ABO, F., NATORI, T., TOTSUKA, T.**: Analysis of air pollutant sorption by plants. (3) Sorption under fumigation with NO_2, O_3 or $NO_2 + O_3$. -
Res. Rep. nat. Inst. environ. Studies 11 (Studies on the Effects of Air Pollutants on Plants and Mechanisms of Phytotoxicity.): 213-224, 1980.

*9114 - **OMER, L.S., SCHLESINGER, W.H.**: Field and greenhouse investigations of the effect of increasing salt stress on the anatomy of *Jaumea carnosa (Asteraceae)*,
a salt marsh species. - Amer. J. Bot. 67: 1455-1465, 1980.

9115 - **ONDOK, J.P.**: Photosynthesis as dependent on stand architecture. Analysis by
means of a model. - In: UNGER, K., STÖCKER, G. (ed.): Biophysikalische Ökologie und Ökosystemforschung. Pp. 227-238. Akademie-Verlag, Berlin 1981.

*9116 - **OOSTERHUIS, D.M., WIEBE, H.H.**: Hydraulic conductivity and osmotic adjustment
in drought acclimated cotton. - Plant Physiol. 65 (Suppl.): 6, 1980.

9117 - **OPARKA, K.J., GATES, P.**: Transport of assimilates in the developing caryopsis
of rice (*Oryza sativa* L.). The pathways of water and assimilated carbon. -
Planta 152: 388-396, 1981.

*9118 - **ORDAS, A.**: Selection for drought resistance in maize. - Genét. Ibér. 30 - 31:
211-223, 1978 - 1979.

*9119 - **ORTON, T.J.**: Comparison of salt tolerance between *Hordeum vulgare* and *H. ju-
batum* in whole plants and callus cultures. - Z. Pflanzenphysiol. 98: 105-118,
1980.

*9120 - **OSBORNE, D.J.**: Senescence in seeds. - In: THIMANN, K.V. (ed.): Senescence in
Plants. Pp. 13-37. CCR Press, Boca Raton 1980.

*9121 - **OSBORNE, D.J., CUMING, A.C.**: Membrane protein and phospholipid turnover in
imbibed dormant embryos of wild oats. - In: LAIDMAN, D.L., WYN JONES, R.G.
(ed.): Recent Advances in the Biochemistry of Cereals. (Phytochem. Soc. Europe
Symp. 16.) Pp. 105-118. Academic Press, London - San Francisco - New York 1979.

*9122 - **OSMOND, C.B., BJÖRKMAN, O., ANDERSON, D.J.**: Physiological Processes in Plant
Ecology. Toward a Synthesis with *Atriplex*. (Ecological Studies Vol. 36.). -
Springer-Verlag, Berlin - Heidelberg - New York 1980.

9123 - **OSUNUBI, O., DAVIES, W.J.**: Root growth and water relations of oak and birch
seedlings. - Oecologia 51: 343-350, 1981.

*9124 - OSTROWSKI, H., FAY, M.F.: Cool season productivity of five tropical grasses in high rainfall area of south-east Queensland. - Trop. Grassland 13: 149-153, 1979.

9125 - O'TOOLE, J.C., MAGULING, M.A.: Greenhouse selection for drought resistance in rice. - Crop Sci. 21: 325-327, 1981.

9126 - O'TOOLE, J.C., MOYA, T.B.: Water deficits and yield in upland rice. - Field Crops Res. 4: 247-259, 1981.

9127 - O'TOOLE, J.C., SOEMARTONO: Evaluation of a simple technique for characterizing rice root systems in relation to drought resistance. - Euphytica 30: 283-290, 1981.

9128 - OUTLAW, W.H., Jr., MAYNE, B.C., ZENGER, V.E., MANCHESTER, J.: Presence of both photosystems in guard cells of *Vicia faba* L. Implications for environmental signal processing. - Plant Physiol. 67: 12-16, 1981.

9129 - OVERDIECK, D., STRAIN, B.R.: Effect of atmospheric humidity on net photosynthesis, transpiration, and stomatal resistance *Hydrocotyle umbellata* L. and *Hydrocotyle bonariensis* Lam. - Int. J. Biometeorol. 25: 29-38, 1981.

*9130 - OVERMAN, A.J., MARTIN, F.G., GREEN, J.L., ENGLEHARD, A.W.: The influence of linear drip irrigation placement in mulched and nonmulched soils on chrysanthemum production and nutrient distribution. - Proc. Florida State hort. Soc. 92: 322-326, 1979.

9131 - ÖZTÜRK, M., REHDER, H., ZIEGLER, H.: Biomass production of C_3 and C_4-plant species in pure and mixed culture with different water supply. - Oecologia 50: 73-81, 1981.

9132 - PAGE, J.M.: Drought-accelerated parasitism of conifers in the Mountain Ranges of Northern California. - Environ. Conserv. 8: 217-226, 1981.

9133 - PALACOIS, E.V.: Response function of crops yield to soil moisture stress. - Water Resour. Bull. 17: 699-703, 1981.

9134 - PALEG, L.G., ASPINALL, D.: The Physiology and Biochemistry of Drought Resistance in Plants. - Academic Press, Sydney - New York - London - Toronto - San Francisco 1981.

9135 - PALEVITZ, B.A.: The structure and development of stomatal cells. - In: JARVIS, P.G., MANSFIELD, T.A. (ed.): Stomatal Physiology. Pp. 1-23. Cambridge University Press, Cambridge - London - New York - New Rochelle - Melbourne - Sydney 1981.

9136 - PALEVITZ, B.A., O'KANE, D.J.: Epifluorescence and video analysis of vacuole motility and development in stomatal cells of *Allium*. - Science 214: 443-445, 1981.

*9137 - PALFI, Z., PALFI, G.: Pokazatel za sukhoustoĭchivost na novi sortove rasteniya ot vidove ot neprolinov tip. [An indicator for the drought resistance of new non-proline type plant cultivars.] - Fiziol. Rast. (Sofia) 5 (3): 43-53, 1979. [In Bulg, ab: R, E.]

9138 - PALIT, P., BHATTACHARYYA, A.C.: Germination and water uptake of jute seeds under water stress. - Ind. J. exp. Biol. 19: 848-852, 1981.

9139 - PALLARDY, S.G.: Closely related woody plants. - In: KOZLOWSKI, T.T. (ed.): Water Deficits and Plants Growth. Volume VI. Woody Plant Communities. Pp. 511-548. Academic Press, New York - San Francisco - London 1981.

*9140 - PALLARDY, S.G., KOZLOWSKI, T.T.: Early root and shoot growth of *Populus* clones. - Silvae Genet. 28: 153-156, 1979.

9141 - PALLARDY, S.G., KOZLOWSKI, T.T.: Water relations of *Populus* clones. - Ecology
62: 159-169, 1981.

*9142 - PALTA, J.P., STADELMANN, E.J.: Simultaneous transport of water and solutes
through plant cell membranes. - In: SPANSWICK, R.M., LUCAS, W.J., DAINTY, J.
(ed.): Plant Membrane Transport: Current Conceptual Issues. Pp. 457-458. Else-
vier/North-Holland Biomedical Press, Amsterdam - New York - Oxford 1980.

9143 - PALTI, J.: Cultural Practices and Infectious Crop Diseases. (Advanced Series
in Agricultural Sciences 9). - Springer-Verlag, Berlin - Heidelberg - New
York 1981.

*9144 - PANDE, P.C.: Foliar epidermis and development of stomata in Iridaceae. - Acta
bot. Ind. 8: 256-259, 1980.

*9145 - PANTOVIČ, S.: Uticaj različite dubine obrade zemljišta na prinos kukuruza na
Kosovskoj Smonici. [Effect of different tilling depths on the yield of maize
on Smonitza of Kosovo.] - Arh. poljoprivredne Nauke 41 (141): 67-89, 1980.
[In Croat, ab: E.]

*9146 - PARASHAR, K.S.: Dry matter production and the concentration and uptake of
nitrogen by sugarbeet as influenced by soil-moisture regime, plant population
and levels of nitrogen. - Ind. J. Agron. 25: 636-640, 1980.

*9147 - PARASHAR, K.S., SARAF, C.S., SHARMA, R.P.: Effect of soil-moisture and ferti-
lizer levels on spring-planted sugarcane grown pure and intercropped with
mung. - Ind. J. Genet. Plant Breed. A40: 103-107, 1980.

9148 - PARKINSON, K.J., DAY, W.: Water vapour calibration using salt hydrate transi-
tions. - J. exp. Bot. 32: 411-418, 1981.

9149 - PARSONS, D.J., RUNDEL, P.W., HEDLUND, R.P., BAKER, G.A.: Survival of severe
drought by a non-sprouting chaparral shrub. - Amer. J. Bot. 68: 973-979, 1981.

9150 - PARSONS, I.T., HUNT, R.: Plant growth analysis: A program for the fitting of
lengthy series of data by the method of *B*-splines. - Ann. Bot. 48: 341-352,
1981.

*9151 - PARSONS, L.R., HOWE, T.K.: Effects of water stress on osmotic and turgor
changes in common beans and the drought resistant tepary bean. - Plant Physiol.
65 (Suppl.): 7, 1980.

9152 - PARTON, W.J., LAUENROTH, W.K., SMITH, F.M.: Water loss from a shortgrass
steppe. - Agr. Meteorol. 24: 97-109, 1981.

9153 - PASSIOURA, J.B.: Water collection by roots. - In: PALEG, L.G., ASPINALL, D.
(ed.): The Physiology and Biochemistry of Drought Resistance in Plants. Pp.
39-53. Academic Press, Sydney - New York - London - Toronto - San Francisco
1981.

9154 - PATE, J.S., LAYZELL, D.B.: Carbon and nitrogen partitioning in the whole plant
- a thesis based on empirical modeling. - In: BEWLEY, J.D. (ed.): Nitrogen and
Carbon Metabolism. Pp. 94-134. Martinus Nijhoff / Dr. W. Junk Publishers, The
Hague - Boston - London 1981.

*9155 - PATERNO, E.S., TILO, S.N.: Influence of soil moisture tension on survival of
rhizobia. - Kalikasan (Philippine J. Biol.) 9 (1): 88-92, 1980.

*9156 - PATHAN, S.N., NIMBALKAR, J.D.: Physiological changes in senescent leaves of *Alternanthera paronychioides* St. Hill and *Alternanthera ficoidea* (L) R. BR. D. - Biovigyanam 5: 143-148, 1979.

*9157 - PATTERSON, D.T.: Light and temperature adaptation. - In: HESKETH, J.D., JONES, J.W. (ed.): Predicting Photosynthesis for Ecosystem Models. Vol. I. Pp. 205-235. CRC Press, Boca Raton 1980.

9158 - PATTERSON, D.T.: Effects of allelopathic chemicals on growth and physiological responses of soybean (*Glycine max*). - Weed Res. 29: 53-59, 1981.

9159 - PEACE, W.J.H., MacDONALD, F.D.: An investigation of the leaf anatomy, foliar mineral levels, and water relations of trees of a Sarawak forest. - Biotropica 13: 100-109, 1981.

9160 - PEARCY, R.W., TUMOSA, N., WILLIAMS, K.: Relationships between growth, photosynthesis and competitive interactions for a C_3 and a C_4 plant. - Oecologia 48: 371-376, 1981.

9161 - PENCHEV, P.N.: Vliyanie na temperaturata na v"zdukha, vlazhnostta na pochvata i sroka na seitba v"rkhu dinamikata na bratene pri zimniya dvureden echemik. [The influence of air temperature, soil moisture and date of seeding on the dynamics of tillering in winter two-row barley.] - Rasteniev. Nauki 18 (4): 60-67, 1981. [In Bulg, ab: R, E.]

9162 - PENOT, M., BÉRAUD, J., PODER, D.: Relationship between hormone-directed transport and transpiration in isolated leaves of *Pelargonium zonale* (L.) Aiton. - Physiol. vég. 19: 391-399, 1981.

9163 - PEREGUDA, L.V., SHELYAG-SOSONKO, Yu.R.: Metodychni aspekty vyvchennya gidrotermichnogo rezhymu roslynnosti. [Methodical aspects of studies on canopy hydrothermal regime.] - Ukr. bot. Zh. 38 (2): 1-8, 1981. [In Ukr, ab: E.]

*9164 - PERRIER, A., KATERJI, N., GOSSE, G., ITIER, B.: Etude "in situ" de l'évapotranspiration réelle d'une culture de blé. - Agr. Meteorol. 21: 295-311, 1980.

*9165 - PERRY, D.A., LOTAN, J.E., HINZ, P., HAMILTON, M.A.: Variation in lodgepole pine: Family response to stress induced by polyethylene glycol 6000. - Forest Sci. 24: 523-526, 1978.

*9166 - PERSSON, T. (ed.): Structure and Function of Northern Coniferous Forests - An Ecosystem Study. (Ecol. Bull. 32.) - Swedish Natural Science Research Council, Stockholm 1980.

*9167 - PERTTU, K., BISCHOF, W., GRIP, H., JANSSON, P.-E., LINDGREN, Å., LINDROTH, A., NORÉN, A.: Micrometeorology and hydrology of pine forest ecosystems. I. Field studies. - Ecol. Bull. (Stockholm) 32 (PERSSON, T. (ed.): Structure and Function of Northern Coniferous Forests - An Ecosystem Study): 75-121, 1980.

9168 - PETERSON, C.A., PETERSON, R.L., BARKER, W.G.: Observations on the structure and osmotic potentials of parenchyma associated with the internal phloem of potato tubers. - Amer. Potato J. 58: 575-584, 1981.

*9169 - PETERSON, C.M., MOLZ, F.J., DANE, J.H.: Effect of chemical treatment on water flow from roots to soil. - In: SPANSWICK, R.M., LUCAS, W.J., DAINTY, J. (ed.): Plant Membrane Transport: Current Conceptual Issues. Pp. 461-462. Elsevier/North-Holland Biomedical Press, Amsterdam - New York - Oxford 1980.

9170 - PETERSON, J.C., SACALIS, J.N., DURKIN, D.J.: *Ficus benjamina*: Avoid water stress to prevent leaf shedding. - Florists' Rev. 1981: 10-11, 37-38, 1981.

*9171 - PETINOV, N.S., KOLESNIK, T.I., KHARANYAN, N.N., KIRILLINA, V.I., EGOROV, V.G.,
       GOLOVATYĬ, V.G.: Svyaz' produktivnosti pshenitsy i nakopleniya belka v zerne
       s urovnem mineral'nogo pitaniya i vlazhnost'yu substrata. [Relationship of
       wheat productivity and grain protein accumulation to mineral nutrition and
       moisture supply.] - Fiziol. Rast. 27: 1277-1287, 1980. [In R, ab: E.]

*9172 - PETKOV, P.S.: Prouchvane deĭstvieto i vzaimodeĭstvieto na posevnata norma,
       sroka na seĭtba i fona na torene v"rkhu produktivnostta na pshenitsata pri
       polivni usloviya v yugoiztochna B"lgariya. [A study on the effect and inter-
       action of seeding rate and seeding date and of fertilizer application on the
       productivity of four wheat cultivars grown under irrigation in south eastern
       Bulgaria.] - Rasteniev. Nauki 17 (2): 55-65, 1980. [In Bulg, ab: E, R.]

9173 - PHELOUNG, P., BARLOW, E.W.R.: Respiration and carbohydrate accumulation in
       the water-stressed wheat apex. - J. exp. Bot. 32: 921-931, 1981.

9174 - PHILLIPS, D.L.: End-point recognition in pressure chamber measurements of wat-
       er potential of *Viguiera porteri* (Asteraceae). - Ann. Bot. 48: 905-907, 1981.

9175 - PICKARD, W.F.: How does the shape of the substomatal chamber affect transpi-
       rational water loss? - Math. Biosci. 56: 111-127, 1981.

*9176 - PIERCE, M., RASCHKE, K.: The role of turgor in the regulation of abscisic acid
       levels in leaves. - Plant Physiol. 65 (Suppl.): 7, 1980.

9177 - PIERCE, M., RASCHKE, K.: Synthesis and metabolism of abscisic acid in detached
       leaves of *Phaseolus vulgaris* L. after loss and recovery of turgor. - Planta
       153: 156-165, 1981.

9178 - PILL, W.G.: Effect of nitrapyrin and nitrate level on growth, elemental compo-
       sition, and water relations of tomato grown in peat-vermiculite. - J. Amer.
       Soc. hort. Sci. 106: 285-289, 1981.

*9179 - PINNERUP, S.P.: Leaf production of *Zostera marina* L. at different salinities.
       - Ophelia 1980 (Suppl. 1): 219-224, 1980.

9180 - PINTÉR, L., KÁLMÁN, L.: Effect of leaf area reduction on standability and
       premature death of leaves in maize (*Zea mays* L.) hybrids with different geno-
       types. - Acta agron. Acad. Sci. Hung. 30: 61-67, 1981.

9181 - PINTÉR, L., PÁLFI, G., KÁLMÁN, L.: Individual analyses of maize (*Zea mays* L.)
       plants for increased drought resistance. - Z. Pflanzenzücht. 87: 260-263, 1981.

*9182 - PITMAN, M.G.: Effect of inhibitors on hydraulic conductivity of roots. - In:
       SPANSWICK, R.M., LUCAS, W.J., DAINTY, J. (ed.): Plant Membrane Transport:
       Current Conceptual Issues. Pp. 463-464. Elsevier/North-Holland Biomedical
       Press, Amsterdam - New York - Oxford 1980.

9183 - PITMAN, M.G.: Ion uptake. - In: PALEG, L.G., ASPINALL, D. (ed.): The Physi-
       ology and Biochemistry of Drought Resistance in Plants. Pp. 71-96. Academic
       Press, Sydney - New York - London - Toronto - San Francisco 1981.

9184 - PITMAN, M.G., WELLFARE, D., CARTER, C.: Reduction of hydraulic conductivity
       during inhibition of exudation from excised maize and barley roots. - Plant
       Physiol. 67: 802-808, 1981.

*9185 - PLAVŠIĆ-GOJKOVIĆ, N., DUBRAVEC, K.: Investigation of cell wall thickenings
       and crystals in the pericarp of *Phaseolus vulgaris* L. - Acta bot. Croat. 38:
       35-40, 1979.

9186 - PLHÁK, F.: Photosynthetic efficiency determined as dry matter increment and
       transpiration rate in alfalfa genotypes. - Photosynthetica 15: 457-466, 1981.

9187 - **POHLMANN, F.W., KUIPER, P.J.C.:** Effect of α-tocopherol on the phase transition of phosphatidylcholine and on water transport through phosphatidylcholine liposome membranes. - Phytochemistry 20: 1525-1528, 1981.

9188 - **POLJAKOFF-MAYBER, A.:** Ultrastructural consequences of drought. - In: PALEG, L.G., ASPINALL, D. (ed.): The Physiology and Biochemistry of Drought Resistance in Plants. Pp. 389-403. Academic Press, Sydney - New York - London - Toronto - San Francisco 1981.

9189 - **POLONENKO, D.R., DUMBROFF, E.B., MAYFIELD, C.I.:** Microbial responses to salt--induced osmotic stress II. Population changes in the rhizoplane and rhizosphere. - Plant Soil 63: 415-426, 1981.

9190 - **POLONENKO, D.R., MAYFIELD, C.I., DUMBROFF, E.B.:** Microbial responses to salt--induced osmotic stress I. Population changes in an agricultural soil. - Plant Soil 59: 269-285, 1981.

9191 - **POLUĖKTOV, R.A.:** Imitatsionnaya model' vlagoobmena v sisteme "pochva - rastenie - atmosfera". [Immitation model of water exchange in the system "soil - plant - atmosphere.] - In: UNGER, K., STÖCKER, G. (ed.): Biophysikalische Ökologie und Ökosystemforschung. Pp. 145-152. Akademie-Verlag, Berlin 1981. [In R.]

9192 - **POLYAKOV, M.A., KARPUSHKIN, L.T.:** Vlazhnost' vozdukha nad isparyayushcheĭ poverkhnost'yu mezofil'nykh kletok lista. Analiz vozmozhnykh prichin snizheniya vlazhnosti. [Air humidity above the evaporating surface of leaf mesophyll cells. Analysis of possible reasons for air humidity decrease.] - Fiziol. Rast. 28: 448-460, 1981. [In R, ab: E.]

*9193 - **POLYAKOV, M.A., MARFENKO, Yu.L., NOVICHKOVA, N.S., VOSKRESENSKAYA, N.P.:** Bystroe uvelichenie CO_2-gazoobmena u zlakov pri deĭstvii sinego sveta. [Rapid increase in CO_2 exchange in cereals under blue light.] - Fiziol. Rast. 27: 1134-1142, 1980. [In R, ab: E.]

9194 - **POOLE, D.K., MILLER, P.C.:** The distribution of plant water stress and vegetation characteristics in southern California chaparral. - Amer. Midl. Natur. 105: 32-43, 1981.

9195 - **POOLE, D.K., ROBERTS, S.W., MILLER, P.C.:** Water utilization. - In: MILLER, P.C. (ed.): Resource Use by Chaparral and Matorral. A Comparison of Vegetation Function in Two Mediterranean Type Ecosystems. Pp. 123-149. Springer-Verlag, New York - Heidelberg - Berlin 1981.

9196 - **POOLE, R.T., CONOVER, C.A.:** Growth response of foliage plants to night and water temperatures. - HortScience 16: 81-82, 1981.

*9197 - **POPOVIĆ, Ž., STIKIĆ, R., PEKIĆ, S.:** Vodni režim kukuruza. [Water regime in maize.] - In: BELIĆ, J. (ed.): Fiziologija Kukuruza. Pp. 111-126. Serb. Acad. Sci. Arts., Beograd 1980. [In Serb, ab: E.]

*9198 - **PORTO, M.C.M., dos SANTOS FILHO, J.M., BARNI, N.A., MINOR, H.C., BERGAMASCHI, H.:** Resposta da soja (*Glycine max* (L.) Merril) à irrigação e níveis de fertilidade do solo. I: Rendimento de grãos e características agronômicas. [Effect of irrigation and nutrient levels on soybean (*Glycine max* (L.) Merrill) yield. I: Grain yield and agronomical characteristics.] - Agron. sulriograndense 15: 251-280, 1979. [In Port, ab: E.]

*9199 - **PORTO, M.C.M., dos SANTOS FILHO, J.M., BARNI, N.A., MINOR, H.C., BERGAMASCHI, H.:** Resposta da soja (*Glycine max* (L.) Merrill) à irrigação e níveis de fertilidade do solo. II: Absorção de nutrientes. [Effect of irrigation and nutrient levels on soybean (*Glycine max* (L.) Merrill) yield. II: Nutrients absorption.] - Agron. sulriograndense 16: 45-56, 1980. [In Port, ab: E.]

9200 - **POSGAY, E.**: A vízellátás és a termés közötti kapcsolat az öntözéses növényter-
        mesztésben. II. Cukorrépa. [Relationship between water supply and yield in
        crop growing under irrigation. II. Sugar beet.] - Növénytermélés 30: 177-188,
        1981. [In Hung, ab: E.]

9201 - **POSGAY, E.**: A vízellátás és a termés közötti kapcsolat az öntözéses növényter-
        mesztésben. III. Burgonya. [Relationship between water supply and yield in
        crop growing under irrigation. III. Potatoes.] - Növenytermelés 30: 353-362,
        1981. [In Hung, ab: E.]

9202 - **POSPÍŠILOVÁ, J., SOLÁROVÁ, J. (ed.)**: Water-in-Plants Bibliography. Volume 5
        1979. - Dr. W. Junk bv Publishers, The Hague - Boston - London 1981.

9203 - **POWELL, A.A., MATTHEWS, S.**: A physical explanation for solute leakage from dry
        pea embryos during imbibition. - J. exp. Bot. 32: 1045-1050, 1981.

9204 - **POWER, J.F., SANDOVAL, F.M., RIES, R.E., MERRILL, S.D.**: Effects of topsoil
        and subsoil thickness on soil water content and crop production on a dis-
        turbed soil. - Soil Sci. Soc. Amer. J. 45: 124-129, 1981.

*9205 - **PRABHA, C., BHARTI, S.**: Effect of ascorbic acid on proline accumulation in
        cowpea leaves under water stress conditions. - Ind. J. Plant Physiol. 23: 317
        - 318, 1980.

*9206 - **PRIEHRADNÝ, S.**: Die Reaktion anfälliger und resistenter Gerstensorten auf
        pilzliche Krankheitserreger. III. Wasserbilanz. - Phytopathol. Z. 98: 218-227,
        1980.

9207 - **PRIEHRADNÝ, S.**: Changes of the water saturation deficit after infection with
        powdery mildew of a susceptible barley cultivar. - Biológia (Bratislava) 36:
        37-42, 1981.

9208 - **PRIHAR, S.S., SANDHU, K.S., KHERA, K.L., SANDHU, B.S.**: Effect of irrigation
        schedules on yield of mustard (*Brassica juncea*). - Exp. Agr. 17: 105-111,
        1981.

9209 - **PROCTOR, J.T.A.**: Stomatal conductance changes in leaves of McIntosh apple
        trees before and after fruit removal. - Can. J. Bot. 59: 50-53, 1981.

*9210 - **PROCTOR, M.C.F.**: Diffusion resistances in bryophytes. - In: GRACE, J., FORD,
        E.D., JARVIS, P.G. (ed.): Plants and their Atmospheric Environment. Pp. 219-
        229. Blackwell Scientific Publications, Oxford - London - Edinburgh - Boston
        - Melbourne 1980.

*9211 - **PROEBSTING, E.L., Jr., MIDDLETON, J.E.**: The behavior of peach and pear trees
        under extreme drought stress. - J. Amer. Soc. hort. Sci. 105: 380-385, 1980.

9212 - **PROEBSTING, E.L., Jr., MIDDLETON, J.E., MAHAN, M.O.**: Performance of bearing
        cherry and prune trees under very low irrigation rates. - J. Amer. Soc. hort.
        Sci. 106: 243-246, 1981.

9213 - **PROKOF'EV, A.A., RYBALOVA, B.A., ZAVADSKAYA, O.Yu.**: Vodoobmen sozrevayushchikh
        plodov maka. [Water exchange in ripening poppy fruit.] - Fiziol. Rast. 28:
        120-129, 1981. [In R, ab: E.]

9214 - **PRUNTY, L.**: Sunflower cultivar performance as influenced by soil water and
        plant population. - Agron. J. 73: 257-260, 1981.

*9215 - **PRZYKORSKA-ŻELAWSKA, T.**: Response of individual Scots pine (*Pinus silvestris*
        L.) seedlings to nutrition and watering as a measure of genotype-environment
        interaction. - Acta Physiol. Plant. 2: 247-255, 1980.

9216 - PRZYKORSKA-ŻELAWSKA, T., ŻELAWSKI, W.: Changes in photosynthetic capacity of non growing needles of conifers during their acclimation to shade. - Acta Physiol. Plant. 3: 33-41, 1981.

9217 - PUCKRIDGE, D.W., O'TOOLE, J.C.: Dry matter and grain production of rice, using a line source sprinkler in drought studies. - Field Crops Res. 3: 303-319, 1981.

9218 - PULKRÁBEK, J., FOREJT, F., VÍTKOVÁ, A.: Vliv vlhkosti půdy na vzcházivost osiva cukrovky. [The effect of soil humidity on the emergence of sugar-beet seed.] - Rost. Výroba (Praha) 27: 1113-1120, 1981. [In Czech, ab: E, G, R.]

9219 - PUSTOVOĬTOVA, T.N., BORODINA, N.A.: Osobennosti adaptatsionnykh reaktsiĭ poliploidnykh rasteniĭ v usloviyakh pochvennoĭ i atmosfernoĭ zasukh. [Some features of adaptive responses of polyploid plants to soil and atmospheric drought.] - Fiziol. Rast. 28: 587-593, 1981. [In R, ab: E.]

9220 - PYATROŬ, Ya.G., SMALYAK, L.P.: Vil'gatsezabyaspechanasts' i praduktsyĭnasts' lyasnykh fitatsenozaŭ na glebakh atmasfernaga ŭvil'gatnennya. [Moisture supply and productivity of forest phytocenoses on the soils with atmospheric moisture supply.] - Vestsi Akad. Navuk Belarus. SSR, Ser, biyal. Navuk 1981 (4): 29-34, 139, 1981. [In Belorus, ab: E, R.]

9221 - QUARRIE, S.A., HENSON, I.E.: Abscisic acid accumulation in detached cereal leaves in response to water stress. II. Effects of leaf age and leaf position. - Z. Pflanzenphysiol. 101: 439-446, 1981.

*9222 - QUIN, B.F., DREWITT, E.G.: The growth, nitrogen uptake and yield of spring-sown wheat on lismore silt loam. - Proc. agron. Soc. N. Zeal. 9: 133-138, 1979.

9223 - RABE, R.: Beeinflussung von physiologischen Prozessen in Pflanzen durch Luftverunreinigungen und ihre Bedeutung für die Stabilität von Ökosystemen. - Angew. Bot. 55: 211-225, 1981.

9224 - RADIN, J.W.: Water relations of cotton plants under nitrogen deficiency. IV. Leaf senescence during drought and its relation to stomatal closure. - Physiol. Plant. 51: 145-149, 1981.

*9225 - RADIN, J.W., ACKERSON, R.C.: Interaction of nitrogen nutrition and water stress in cotton: Control of stomatal behavior and gas exchange. - Plant Physiol. 65 (Suppl.): 74, 1980.

9226 - RADIN, J.W., ACKERSON, R.C.: Water relations of cotton plants under nitrogen deficiency. III. Stomatal conductance, photosynthesis, and abscisic acid accumulation during drought. - Plant Physiol. 67: 115-119, 1981.

9227 - RADLEY, M.: The effect on wheat grain growth of the removal or ABA treatment of glumes and lemmas. - J. exp. Bot. 32: 129-140, 1981.

*9228 - RADOSEVICH, S.R., CONARD, S.G.: Physiological control of chamise shoot growth after fire. - Amer. J. Bot. 67: 1442-1447, 1980.

*9229 - RADOSEVICH, S.R., RONCORONI, E.J., CONARD, S.G., MCHENRY, W.B.: Seasonal tolerance of six coniferous species to eight foliage-active herbicides. - Forest Sci. 26: 3-9, 1980.

*9230 - RAGHAVENDRA, A.S.: Chloride and nitrate stimulation of stomatal opening in relation to phosphoenolpyruvate carboxylation and malate production in epidermal tissues of *Commelina benghalensis*. - Plant Physiol. 65 (Suppl.): 49, 1980.

*9231 - **RAGHAVENDRA, A.S.:** Energy supply for stomatal opening in epidermal strips of
         Commelina benghalensis: Primary source of oxidative phosphorylation and sup-
         plementary role of photophosphorylation. - Plant Physiol. 65 (Suppl.): 49,
         1980.

 9232 - **RAGHAVENDRA, A.S.:** Energy supply for stomatal opening in epidermal strips of
         Commelina benghalensis. - Plant Physiol. 67: 385-387, 1981.

 9233 - **RAHMAN, S.M., TALUKDAR, S.U., KAUL, A.K., BISWAS, M.R.:** Yield response of a
         semi-dwarf wheat variety to irrigation on a calcareous brown flood plain soil
         of Bangladesh. - Agr. Water Manage. 3: 217-225, 1981.

 9234 - **RAI, A.K., PANDEY, G.P.:** Influence of environmental stress on the germination
         of *Anabaena vaginicola* akinetes. - Ann. Bot. 48: 361-370, 1981.

*9235 - **RAI, R.S.V., MURTY, K.S.:** Growth and yield in rice as influenced by higher
         nitrogen application under normal and water-logged conditions. - Ind. J. agr.
         Res. 13: 133-137, 1979.

*9236 - **RAJ, K.P.S., CHINOY, J.J.:** RNase activity and nucleic acids content as affec-
         ted by drought and photocycles in barley varieties. - Fyton 39: 77-84, 1980.

 9237 - **RAJAGOPAL, V.:** The influence of exogenous proline on the stomatal resistance
         in *Vicia faba*. - Physiol. Plant. 52: 292-296, 1981.

 9238 - **RAMANI, S., KANNAN, S., NIRALE, A.S.:** Differential zinc uptake and transport
         in sorghum varieties suited for different moisture regimes. - J. Plant Nutr.
         4: 337-353, 1981.

 9239 - **RAO, A.N., DAS, V.S.R.:** Leaf photosynthetic characters and crop growth rate
         in six cultivars of groundnut (*Arachis hypogaea* L.) - Photosynthetica 15: 97-
         103, 1981.

*9240 - **RAO, A.S., NAINAWATEE, H.S.:** Water-stress associated proline accumulation in
         wheat seedlings. - Haryana agr. Univ. J. Res. 10: 365-368, 1980.

 9241 - **RAO, G.G., BASHA, S.K.M., RAO, G.R.:** Effect of NaCl salinity on amount and
         composition of epicuticular wax and cuticular transpiration rate in peanut
         Arachis hypogaea L. - Ind. J. exp. Biol. 19: 880-881, 1981.

 9242 - **RAO, G.G., RAO, G.R.:** Pigment composition & chlorophyllase activity in pigeon
         pea (*Cajanus indicus* Spreng) and gingelley (*Sesamum indicum* L.) under NaCl
         salinity. - Ind. J. exp. Biol. 19: 768-770, 1981.

*9243 - **RAO, J.V.S., MURTHY, S.S., REDDY, K.R.:** The quality of leaf epicuticular wax
         in relation to the rates of transpiration of some angiospermous weed species.
         - Comp. Physiol. Ecol. 5: 175-179, 1980.

*9244 - **RAO, J.V.S., REDDY, K.R.:** Seasonal variation in leaf epicuticular wax of some
         semiarid shrubs. - Ind. J. exp. Biol. 18: 495-499, 1980.

 9245 - **RASMUSSEN, H.:** Terminology and classification of stomata and stomatal develop-
         ment- a critical survey. - Bot. J. Linn. Soc. 83: 199-212, 1981.

 9246 - **RASMUSSEN, H.:** The diversity of stomatal development in *Orchidaceae* subfamily
         Orchidoideae. - Bot. J. Linn. Soc. 82: 381-393, 1981.

 9247 - **RATHFELDER, E.L.:** The effect of different water regimes on plant growth and
         nodule structure of greenhouse-grown winged bean (*Psophocarpus tetragonolobus*).
         - Ann. appl. Biol. 98: 143-148, 1981.

9248 - RATKOVIĆ, S., BAČIĆ, G.: NMR study of water in plant cells and tissues. Proton T_1 and hydration level of *Zea mays* root in relation to ion transport. - Studia biophys. 84: 77-78, 1981.

*9249 - RAUSER, W.E., DUMBROFF, E.B.: Water relations of bean seedlings exposed to excess cobalt, nickel, and zinc. - Plant Physiol. 65 (Suppl.): 85, 1980.

9250 - RAUSER, W.E., DUMBROFF, E.B.: Effects of excess cobalt, nickel and zinc on the water relations of *Phaseolus vulgaris*. - Environ. exp. Bot. 21: 249-255, 1981.

9251 - RAVINDRAN, A., ASHRAF, J., AMIN, M.: Relevance of the membrane resting potential for the study of drought resistance in wheat (*Triticum aestivum* L.) cultivars. - Curr. Sci. 50: 908-910, 1981.

9252 - RAWLINS, S.L.: Principles of salinity control in irrigated agriculture. - In: MANASSAH, J.T., BRISKEY, E.J. (ed.): Advances in Food-Producing Systems for Arid and Semiarid Lands. Part A. Pp. 391-420. Academic Press, New York - San Francisco - London 1981.

9253 - RAYDER, L., TING, I.P.: Carbon metabolism in two species of *Pereskia* (Cactaceae). - Plant Physiol. 68: 139-142, 1981.

*9254 - REDDY, K.R., RAO, J.V.S., DAS, V.S.R.: Inhibition of epicuticular wax deposition by ansar 529 & EPTC in some semiarid shrubs. - Ind. J. exp. Biol. 17: 1-3, 1979.

9255 - REDDY, P.K.R., SHAH, G.L.: On the structure, ontogeny and distribution of stomata and trichomes on the pericarps of *Cassia sophera* L. - Bangladesh J. Bot. 10: 43-52, 1981.

9256 - REED, R.H.: Osmotic adaptation in *Porphyra purpurea*: Use of $^{86}Rb^+$ and $^{42}K^+$ exchange kinetics to investigate intracellular solute compartmentation. - Plant Sci. Lett. 23: 215-221, 1981.

*9257 - REED, R.H., COLLINS, J.C.: The ionic relations of *Porphyra purpurea* (Roth) C. Ag. (Rhodophyta, Bangiales). - Plant Cell Environ. 3: 399-407, 1980.

*9258 - REED, R.H., COLLINS, J.C., RUSSELL, G.: The effects of salinity upon cellular volume of the marine red alga *Porphyra purpurea* (Roth). C. Ag. - J. exp. Bot. 31: 1521-1537, 1980.

*9259 - REED, R.H., COLLINS, J.C., RUSSELL, G.: The effects of salinity upon galactosyl-glycerol content and concentration of the marine red alga *Porphyra purpurea* (Roth) C. Ag. - J. exp. Bot. 31: 1539-1554, 1980.

9260 - REED, R.H., COLLINS, J.C., RUSSELL, G.: The effects of salinity upon ion content and ion transport of the marine red alga *Porphyra purpurea* (Roth) C. Ag. - J. exp. Bot. 32: 347-367, 1981.

9261 - REES, D.J., GRACE, J.: Effect of wind and shaking on the water relations of *Pinus contorta*. - Physiol. Plant. 51: 222-228, 1981.

9262 - REGGIARDO, L., MORANDI, E.N., NAKAYAMA, F.: Influencia del CCC [cloruro de (2-cloroetil)trimetilamonio] y de distintos niveles hídricos sobre el consumo de agua en soja, *Glycine max* (L.) Merrill. [The effect of CCC (2-chloroethyl) trimethylammonium chloride and different soil water content on soybean, *Glycine max* (L.) Merrill water consumption.] - Fyton 40: 1-11, 1981. [In Span, ab: E.]

*9263 - REICHMAN, G.A., DOERING, E.J., BENZ, L.C., FOLLETT, R.F.: Construction and performance of large automatic (nonweighing) lysimeters. - Trans. ASAE 22: 1343-1346, 1352, 1979.

9264 - REICOSKY, D.C., VOORHEES, W.B., RADKE, J.K.: Unsaturated water flow through a simulated wheel track. - Soil Sci. Soc. Amer. J. 45: 3-8, 1981.

9265 - REID, R.K., PINCKARD, J.A.: Changes in the cellular structure of the cotton peduncle related to water transport and boll opening. - Crop Sci. 21: 717-720, 1981.

*9266 - RENGER, M., STREBEL, O.: Water and nutrient transport to plant roots as a function of depth and time under field conditions. - In: Soils in Mediterranean Type Climates and their Yield Potential. Pp. 79-91. International Potash Institute, Bern 1979.

9267 - RESENDE, M., HENDERSON, D.W., FERERES, E.: Frequência de irrigação, desenvolvimento e produção do feijão kidney. [Irrigation frequency, crop development and yield of kidney bean.] - Pesq. agropec. Brasil. 16: 363-370, 1981. [In: Port, ab: E.]

*9268 - REUTSKIĬ, V.G., VASIL'EV, L.L.: O mekhanizme regulyatsii temperatury rastitel'noĭ tkani pri otsutstvii transpiratsii. [On the mechanism of temperature regulation in plant tissue without transpiration.] - Dokl. Akad. Nauk Belorus. SSR 24: 1033-1036, 1980. [In R, ab: E.]

9269 - REYNOLDS, J.F., CUNNINGHAM, G.L.: Validation of a primary production model of the desert shrub Larrea tridentata using soil-moisture augmentation experiments. - Oecologia 51: 357-363, 1981.

*9270 - RICCI, F.P., SIGNORELLI, G., GIANCARLINI, G., COSTANTINI, A., TOMBESI, L.: Studies on the energetic balance of Class A Pan evaporimeter. - Ann. Ist. sperim. Nutr. Piante 10 (1): 1-24, 1980.

9271 - RICHARDS, R.A., PASSIOURA, J.B.: Seminal root morphology and water use of wheat. I. Environmental effects. - Crop Sci. 21: 249-252, 1981.

9272 - RICHARDS, R.A., PASSIOURA, J.B.: Seminal root morphology and water use of wheat. II. Genetic variation. - Crop Sci. 21: 253-255, 1981.

*9273 - RICHTER, H., DUHME, F., GLATZEL, G., HINCKLEY, T.M., KARLIC, H.: Some limitations and applications of the pressure-volume curve technique in ecophysiological research. - In: GRACE, J., FORD, E.D., JARVIS, P.G. (ed.): Plants and their Atmospheric Environment. Pp. 263-272. Blackwell Scientific Publications, Oxford - London - Edinburgh - Boston - Melbourne 1980.

9274 - RIEBESELL, J.F.: Photosynthetic adaptations in bog and alpine populations of Ledum groenlandicum. - Ecology 62: 579-586, 1981.

9275 - RIES, R.E., POWER, J.F.: Increased soil water storage and herbage production from snow catch in North Dakota. - J. Range Manage 34: 485-488, 1981.

9276 - RINGEL, B.: Einsatz einer einfachen Giessanlage für die Gefässanzucht von Kartoffeln bei unterschiedlicher Bodenfeuchte. - Arch. Acker- Pflanzenbau Bodenk. 24: 655-657, 1981.

9277 - RISSER, G., PITRAT, M., LECOQ, H., RODE, J.-C.: Sensibilité variétale du melon (Cucumis melo L.) au virus du rabougrissement jaune du melon (MYSV) et à sa transmission par Aphis gossypii. Hérédité de la réaction de flétrissement. - Agronomie 1: 835-838, 1981.

9278 - RITCHIE, J.T.: Water dynamics in the soil-plant-atmosphere system. - Plant Soil 58: 81-96, 1981. Also in: MONTEITH, J., WEBB, C. (ed.): Soil Water and Nitrogen in Mediterranean-type Environments. Development in Plant and Soil Sciences, Volume 1. Pp. 81-96. Martinus Nijhoff / Dr. W. Junk Publishers, The Hague - Boston - London 1981.

9279 - **ROBERTSON, W.K., HAMMOND, L.C., PRINE, G.M., MARTIN, F.G.:** Response of corn cultivars on sandy soil to irrigation, row-spacing, plant population, and nitrogen. - Soil Crop Sci. Soc. Florida Proc. 40: 101-105, 1981.

*9280 - **ROBICHAUX, R.H., PEARCY, R.W.:** Environmental characteristics, field water relations, and photosynthetic responses of C_4 Hawaiian *Euphorbia* species from contrasting habitats. - Oecologia 47: 99-105, 1980.

*9281 - **ROBICHAUX, R.H., PEARCY, R.W.:** Photosynthetic responses of C_3 and C_4 species from cool shaded habitats in Hawaii. - Oecologia 47: 106-109, 1980.

*9282 - **ROBINSON, F.E., TANJI, K.K., LUTHIN, J.N., LEHMAN, W.F., MAYBERRY, K.S., SCHNAGL, R.J., PADGETT, W.:** Irrigation management of Colorado River water with increase in salinity. - Trans. ASAE 23: 859-865, 1980.

*9283 - **RODRIGUES, J.D., KLAR, A.E., PEDRAS, J.F., RODRIGUES, S.D., de PINHO, S.Z.:** A influência de diferentes regimes de umidade do solo em gladíolos. I. Transpiração, teor relativo de água e índice refratométrico. [Influence of different soil-humidity regimes on gladiolas. I. Transpiration, water content and refractive index.] - Fyton 39: 57-76, 1980. [In Port, ab: E.]

9284 - **ROGAN, P.K., ZACCAI, G.:** Hydration in purple membrane as a function of relative humidity. - J. mol. Biol. 145: 281-283, 1981.

9285 - **ROGERS, C., SHARPE, P.J.H., POWELL, R.D., SPENCE, R.D.:** High-temperature disruption of guard cells of *Vicia faba*. Effect on stomatal aperture. - Plant Physiol. 67: 193-196, 1981.

*9286 - **RONA, J.-P., PITMAN, M.G., LÜTTGE, U., BALL, E.:** Electrochemical data on compartmentation into cell wall, cytoplasm, and vacuole of leaf cells in the CAM genus *Kalanchoë*.- J. Membrane Biol. 57: 25-35, 1980.

9287 - **ROOS, E.E., STANWOOD, P.C.:** Effects of low temperature, cooling rate, and moisture content on seed germination of lettuce. - J. Amer. Soc. hort. Sci. 106: 30-34, 1981.

9288 - **ROSE, D.A.:** Gas exchange in leaves. - In: ROSE, D.A., CHARLES-EDWARDS, D.A. (ed.): Mathematics and Plant Physiology. Pp. 67-78. Academic Press, London - New York - Toronto - Sydney - San Francisco 1981.

9289 - **ROSE, D.A., CHARLES-EDWARDS, D.A. (ed.):** Mathematics and Plant Physiology. (Experimental Botany. An International Series of Monographs. Volume 16.) - Academic Press, London - New York - Toronto - Sydney - San Francisco 1981.

9290 - **ROSEMA, A.:** Thermal sensing of soil moisture, evaporation and crop yield. - In: BERG, A. (ed.): Application of Remote Sensing to Agricultural Production Forecasting. Pp. 213-223. A.A. Balkema, Rotterdam 1981.

9291 - **ROSIELLE, A.A., HAMBLIN, J.:** Theoretical aspects of selection for yield in stress and non-stress environments. - Crop Sci. 21: 943-946, 1981.

9292 - **ROSSIELLO, R.O.P., FERNANDES, M.S., FLORES, J.P.O.:** Efeitos da deficiência hídriga sobre o crescimento e a acumulação de carboidratos solúveis de milho. [Effects of water stress on the accumulation of soluble carbohydrates and growth patterns of corn.] - Pesq. agropec. Brasil. 16: 561-566, 1981. [In Port, ab: E.]

9293 - **ROSSIELLO, R.P., FERNANDES, M.S., MAZUR, N.:** Efectos del desecamiento del suelo sobre el metabolismo de nitrogeno en tres cultivares de maiz (*Zea mays* L.). [The effects of water stress on the nitrogen metabolism of three corn (*Zea mays* L.) cultivars.] - Turrialba 31: 227-235, 1981. [In Span, ab: E.]

9294 - ROTH-BEJERANO, N., ITAI, C.: Involvement of phytochrome in stomatal movement: Effect of blue and red light. - Physiol. Plant. 52: 201,206, 1981.

9295 - ROTH-BEJERANO, N., ITAI, C.: Effect of boron on stomatal opening in epidermal strips of *Commelina communis.* - Physiol. Plant. 52: 302-304, 1981.

9296 - ROWSE, H.R., GOODMAN, D.: Axial resistance to water movement in broad bean (*Vicia faba*) roots. - J. exp. Bot. 32: 591-598, 1981.

9297 - ROWSE, H.R., STONE, D.A.: Deep cultivation of a sandy clay loam. II. Effects on soil hydraulic properties and on root growth, water extraction and water stress in 1977, especially of broad beans. - Soil Tillage Res. 1: 173-185, 1980/1981.

9298 - RUBAN, M.M.: Praduktsyĭnasts' sfagnavykh imkhoŭ na aligatrofnykh i mezatrof-nykh balotakh Prypyatskaga zapavednika. [Sphagnum moss productivity on oligo-trophic and mesotrophic bogs of the Pripyat reservation.] - Vestsi Akad. Na-vuk Belarus. SSR, Ser. biyal. Navuk 1981 (4): 20-24, 139, 1981. [In Belorus, ab: E, R.]

9299 - RUBINSTEIN, B., CLELAND, R.E.: Response of *Avena* coleoptiles to suboptimal fusicoccin: Kinetics and comparisons with indoleacetic acid. - Plant Physiol. 68: 543-547, 1981.

9300 - RUDICH, J., RENDON-POBLETE, E., STEVENS, M.A., AMBRI, A.-I.: Use of leaf water potential to determine water stress in field-grown tomato plants. - J. Amer. Soc. hort. Sci. 106: 732-736, 1981.

9301 - RÜEGG, J.: Effects of temperature and water stress on the growth and yield of winged bean (*Psophocarpus tetragonolus* (L.) DC.) - J. hort. Sci. 56: 331-338, 1981.

9302 - RUGENSTEIN, S.R., LERSTEN, N.R.: Stomata in seeds and fruits of *Bauhinia* (Le-guminosae: Caesalpinioideae). - Amer. J. Bot. 68: 873-876, 1981.

9303 - RUMBAUGH, M.D., JOHNSON, D.A.: Screening alfalfa germplasms for seedling drought resistance. - Crop Sci. 21: 709-713, 1981.

9304 - RUMI, C.P., CARPINETTI, R.M.: Effect of sunlight on the development of *Tro-paeolum majus* L. II. Leaf development. - Phyton 41: 129-137, 1981.

9305 - RUSH, D.W., EPSTEIN, E.: Comparative studies on the sodium, potassium, and chloride relations of a wild halophytic and a domestic salt-sensitive tomato species. - Plant Physiol. 68: 1308-1313, 1981.

9306 - RUSHIN, J.W., ANDERSON, J.E.: An examination of the leaf quaking adaptation and stomatal distribution in *Populus tremuloides* Michx. - Plant Physiol. 67: 1264-1266, 1981.

9307 - RUSSELLE, M.P., DEIBERT, E.J., HAUCK, R.D., STEVANOVIC, M., OLSON, R.A.: Ef-fect of water and nitrogen management on yield and ^{15}N-depleted fertilizer use efficiency of irrigated corn. - Soil Sci. Soc. Amer. J. 45: 553-558, 1981.

*9308 - RYBERG, H., LILJENBERG, C., SUNDQVIST, C.: Crystalloid formation in proto-chlorophyll-accumulating plastids from the inner seed coat of *Cyclanthera explodens.* - Physiol. Plant. 50: 333-339, 1980.

9309 - RYBKINA, G.V., BIGLOVA, S.G., KOMPANIÉTS', I.I.: Khloroplasty - osnovni skhovishcha ta zapasni rezervuary vody v klityni v umovakh vodnogo stresu. [Chloroplasts as the main collectors and reservoirs of water in cell under conditions of water stress.] - Ukr. bot. Zh. 38 (1): 71-76, 112, 1981. [In Ukr, ab: E, R.]

*9310 - **RYBKINA, G.V., BIGLOVA, S.G., PAL'M, G.G.:** O kharaktere ob"emnykh izmeneniĭ khloroplastov pri izmenenii osmoticheskogo potentsiala sredy. [Character of volume changes in chloroplasts during changing of osmotic potential of the medium.] - Uch. Zapiski Kazan. gos. pedagog. Inst. 195 (Faktory Sredy i Rastenie.): 40-50, 1979. [In R.]

*9311 - **RYBKINA, G.V., LOSEVA, N.L., BIGLOVA, S.G., PAL'M, G.G.:** K izucheniyu korrelyatsiĭ vodnogo balansa i fotokhimicheskoĭ aktivnosti khloroplastov. [Correlations of water balance and photochemical activity of chloroplasts.] - Uch. Zapiski Kazan. gos. pedagog. Inst. 195 (Faktory Sredy i Rastenie.): 30-39, 1979. [In R.]

9312 - **SAFTNER, R.A., RASCHKE, K.:** Electrical potentials in stomatal complexes. - Plant Physiol. 67: 1124-1132, 1981.

*9313 - **SAHA, A.K., CHAUDHAN, C.P.S., YADAV, D.S.:** Studies on soil-water behaviour and crop production under rainfed conditions. - J. Ind. Soc. Soil Sci. 28: 277-285, 1980.

9314 - **SAHA, P.K., TAKAHASHI, N.:** Seed dormancy and water uptake in *Crotalaria sericea* Retz. - Ann. Bot. 47: 423-425, 1981.

*9315 - **SAINI, G.R., GRANT, W.J.:** Long-term effects of intensive cultivation on soil quality in the potato-growing areas of New Brunswick (Canada) and Maine (USA). - Can. J. Soil Sci. 60: 421-428, 1980.

9316 - **SAINI, H.S., ASPINALL, D.:** Effect of water deficit on sporogenesis in wheat (*Triticum aestivum* L.) - Ann. Bot. 48: 623-633, 1981.

9317 - **SAINI, H.S., SRIVASTAVA, A.K.:** Osmotic stress and the nitrogen metabolism of two groundnut (*Arachis Hypogaea* L.) cultivars. - Irrig. Sci. 2: 185-192, 1981.

*9318 - **SAINT-MARTIN, M.:** Types stomatiques des plantules de *Papilionacées*. - Bull. Soc. bot. France 126 (Actual. bot. 3): 85-91, 1979.

9319 - **SAKAMOTO, C.M.:** Climate-crop regression yield model: An appraisal. - In: BERG, A. (ed.): Application of Remote Sensing to Agricultural Production Forecasting. Pp. 131-138. A.A. Balkema, Rotterdam 1981.

*9320 - **SAKSON, N.:** Wplyw stałej i zmiennej wilgotnosci podłoża na plonowanie pokrzyku wilczej jagody (*Atropa belladonna* L.). [Influence of constant and changing ground moisture on the cropping of *Atropa belladonna* L.] - Herba Pol. 26: 161-165, 1980. [In Pol, ab: E, R.]

9321 - **SALA, O.E., LAUENROTH, W.K., PARTON, W.J., TRLICA, M.J.:** Water status of soil and vegetation in a short-grass steppe. - Oecologia 48: 327-331, 1981.

9322 - **SALAMA, F.M., KHODARY, S.E.A., HEIKAL, M.M.:** Effects of soil salinity and IAA on growth, photosynthetic pigments, and mineral composition of tomato and rocket plants. - Phyton 21: 177-188, 1981.

9323 - **SAMIEV, Kh.S., MARFINA, K.G.:** Osobennosti belkov membrannykh fraktsiĭ khloroplastov u raznykh po zasukhoustoĭchivosti sortov khlopchatnika pri vodnom defitsite. [Peculiarities of chloroplast membrane fraction proteins in different drought resistant cotton cultivars under moisture deficit.] - Sel'skokhoz. Biol. 16: 834-835, 1981. [In R, ab: E.]

9324 - **SAMMIS, T.W.:** Yield of alfalfa and cotton as influenced by irrigation. - Agron. J. 73: 323-329, 1981.

9325 - **SAMMONS, D.J., PETERS, D.B., HYMOWITZ, T.:** Screening soybeans for tolerance to moisture stress: a field procedure. - Field Crops Res. 3: 321-335, 1980/1981.

9326 - **SAMUI, R.P., KAR, S.:** Soil and plant resistance effects on transpirational and leaf water responses by groundnut to soil water potential. - Aust. J. Soil Res. 19: 51-60, 1981.

*9327 - **SANCES, F.V., WYMAN, J.A., TING, I.P.:** Physiological responses to spider mite infestation of strawberries. - Environ. Entomol. 8: 711-714, 1979.

9328 - **SANCES, F.V., WYMAN, J.A., TING, I.P., VAN STEENWYK, R.A., OATMAN, E.R.:** Spider mite interactions with photosynthesis, transpiration and productivity of strawberry. - Environ. Entomol. 10: 442-448, 1981.

9329 - **SANDANAM, S., GEE, G.W., MAPA, R.B.:** Leaf water diffusion resistance in clonal tea (*Camellia sinensis* L.): Effects of water stress, leaf age and clones. - Ann. Bot. 47: 339-349, 1981.

9330 - **SANDERS, D.:** Physiological control of chloride transport in *Chara corallina*. I. Effects of low temperature, cell turgor pressure, and anions. - Plant Physiol. 67: 1113-1118, 1981.

9331 - **SANDERS, D.:** Physiological control of chloride transport in *Chara corallina*. II. The role of chloride as a vacuolar osmoticum. - Plant Physiol. 68: 401-406, 1981.

*9332   **SANTARIUS, K.A., HEBER, U., KRAUSE, G.H.:** Untersuchungen über die physiologisch-biochemischen Ursachen von Empfindlichkeit und Resistenz von Biomembranen gegenüber extremen Temperaturen und hohen Salzkonzentrationen. - Ber. Deut. bot. Ges. 92: 209-223, 1979.

*9333 - **SARKADI, J., PUSZTAI, A.:** Adatok a kis tenyészedények öntözésének módszertanához. [Some data about the method of watering small pots.] - Agrokem. Talajtan 29: 497-510, 1980. [In Hung, ab: E, G, R.]

*9334 - **SATLER, S.O., THIMANN, K.V.:** The influence of aliphatic alcohols on leaf senescence. - Plant Physiol. 66: 395-399, 1980.

9335 - **SATOH, K.:** Fluorescence induction and activity of ferredoxin-NADP$^+$ reductase in *Bryopsis* chloroplasts. - Biochim. biophys. Acta 638: 327-333, 1981.

*9336 - **SAWANO, M.:** [Effects of soil moisture and root temperature on freezing tolerance of chestnut trees.] - Sci. Rep. Fac. Agr. Kobe Univ. 14: 31-35, 1980. [In Jap, ab: E.]

9337 - **SAXE, H., RAJAGOPAL, R.:** Effect of vanadate on bean leaf movement, stomatal conductance, barley leaf unrolling, respiration, and phosphatase activity. - Plant Physiol. 68: 880-884, 1981.

9338 - **SCHÄFER, W.:** Erfassung und Darstellung der Produktivität von Pflanzenbeständen in Abhängigkeit vom Boden-Wasserhaushalt. - In: UNGER, K., STÖCKER, G. (ed.): Biophysikalische Ökologie und Ökosystemforschung. Pp. 137-143. Akademie-Verlag, Berlin 1981.

9339 - **SCHALITZ, G., BREUNIG, W., GROSSKOPF, M., RICHTER, K.:** Leistungsfähigkeit von Luzernebeständen in Abhängigkeit von Nutzungsintensität und Zusatzregengabe. - Arch. Acker- Pflanzenbau Bodenk 25: 311-317, 1981.

9340 - **SCHEURICH, P., ZIMMERMANN, U., SCHNABL, H.:** Electrically stimulated fusion of different plant cell protoplasts. Mesophyll cell and guard cell protoplasts of *Vicia faba*. - Plant Physiol. 67: 849-853, 1981.

9341 - **SCHINNINGER, R.:** Der Einfluss isolierter und kombinierten Schadstoffe auf Austrocknungsresistenz und Transpiration bei *Amaranthus chlorostachys* Willd. - Flora 171: 187-198, 1981.

9342 - **SCHINNINGER, R.**: Der Einfluss isolierter und kombinierter Schadstoffe auf Austrocknungsresistenz und Transpiration bei *Trifolium repens* L. - Phyton 21: 245-259, 1981.

9343 - **SCHLEIFF, U.**: Osmotic potentials of roots of onions and their rhizospheric soil solutions when irrigated with saline drainage waters. - Agr. Water Manage. 3: 317-323, 1981.

9344 - **SCHLEIFF, U.**: Bestimmung der Gehalte an osmotisch wirksamen Substanzen in Pflanzen. - Z. Pflanzenernähr. Bodenk. 144: 335-338, 1980/1981.

9345 - **SCHMIDT, H.W., MÉRIDA, T., SCHÖNHERR, J.**: Water permeability and fine structure of cuticular membranes isolated enzymatically from leaves of *Clivia miniata* Reg. - Z. Pflanzenphysiol. 105: 41-51, 1981.

*9346 - **SCHMITZ, K., SRIVASTAVA, L.M.**: Long distance transport in *Macrocystis integrifolia*. III. Movement of THO. - Plant Physiol. 66: 66-69, 1980.

*9347 - **SCHNABL, H.**: Rapid gluconeogenesis in starch-containing guard cell protoplasts. - In: SPANSWICK, R.M., LUCAS, W.J., DAINTY, J. (ed.): Plant Membrane Transport: Current Conceptual Issues. Pp. 455-456. Elsevier/North-Holland Biomedical Press, Amsterdam - New York - Oxford 1980.

9348 - **SCHNABL, H.**: The compartmentation of carboxylating and decarboxylating enzymes in guard cell protoplasts. - Planta 152: 307-313, 1981.

9349 - **SCHNEPF, E.**: Special cytology: differentiated cells and cell development in higher plants. - Prog. Bot. 43: 13-26, 1981.

*9350 - **SCHOBERT, B.**: Proline catabolism, relaxation of osmotic strain and membrane permeability in the diatom *Phaeodactylum tricornutum*. - Physiol. Plant. 50: 37 -42, 1980.

*9351 - **SCHOBERT, B.**: The function of proline accumulation in the diatom *Phaeodactylum tricornutum* during water stress. - In: SPANSWICK, R.M., LUCAS, W.J., DAINTY, J. (ed.): Plant Membrane Transport: Current Conceptual Issues. Pp. 487-488. Elsevier/North-Holland Biomedical Press, Amsterdam - New York - Oxford 1980.

*9352 - **SCHOENEWEISS, D.F.**: Protection against stress predisposition to *Botryosphaeria canker* in continerized *Cornus stolonifera* by soil injection with benomyl. - Plant Dis. Rep. 63: 896-900, 1979.

9353 - **SCHONBECK, M.W., BEWLEY, J.D.**: Responses of the moss *Tortula ruralis* to desiccation treatments. I. Effects of minimum water content and rates of dehydration and rehydration. - Can. J. Bot. 59: 2698-2706, 1981.

9354 - **SCHONBECK, M.W., BEWLEY, J.D.**: Responses of the moss *Tortula ruralis* to desiccation treatments. II. Variations in desiccation tolerance. - Can. J. Bot. 59: 2707-2712, 1981.

9355 - **SCHÖNHERR, J., LENDZIAN, K.**: A simple and inexpensive method of measuring water permeability of isolated plant cuticular membranes. - Z. Pflanzenphysiol. 102: 321-327, 1981.

9356 - **SCHÖNHERR, J., MÉRIDA, T.**: Water permeability of plant cuticular membranes: the effects of humidity and temperature on the permeability of non-isolated cuticles of onion bulb scales. - Plant Cell Environ. 4: 349-354, 1981.

*9357 - **SCHREMPF, M.**: The action of abscisic acid on the circadian petal movement of *Kalanchoe blossfeldiana*. - Z. Pflanzenphysiol. 100: 397-407, 1980.

9358 - SCHUCK, H.J.: Untersuchungen über die Wasserleitung in am Tannensterben er-
       krankten Weisstannen (*Abies alba* Mill.). - Forstwiss. Centralbl. 100: 184-189,
       1981.

9359 - SCHULT, S., DÖRFFLING, K.: Evidence against an intermediary role of abscisic
       acid in stomatal closure induced by phenylmercuric acetate and farnesol. -
       Physiol. Plant. 53: 487-490, 1981.

9360 - SCHULZE, E.-D., HALL, A.E.: Short-term and long-term effects of drought on
       steady-state and time integrated plant processes. - In: JOHNSON, C.B. (ed.):
       Physiological Processes Limiting Plant Productivity. Pp. 217-235. Butterworths,
       London - Boston - Sydney - Wellington - Durban - Toronto 1981.

9361 - SCHWARZ, M., GALE, J.: Maintenance respiration and carbon balance of plants at
       low levels of sodium chloride salinity. - J. exp. Bot. 32: 933-941, 1981.

9362 - SCHWEIGER, H.G. (ed.): International Cell Biology 1980 - 1981. Papers presented
       at the Second International Congress on Cell Biology, Berlin (West), August
       31 - September 5, 1980. - Springer-Verlag, Berlin - Heidelberg - New York 1981.

9363 - SEEWALDT, V., PRIESTLEY, D.A., LEOPOLD, A.C., FEIGENSON, G.W., GOODSAID-ZAL-
       DUONDO, F.: Membrane organization in soybean seeds during hydration. - Planta
       152: 19-23, 1981.

9364 - SEGUIN, B.: Bioclimatological aspects of crop production. - In: BERG, A. (ed.):
       Application of Remote Sensing to Agricultural Production Forecasting. Pp. 33-
       45. A.A. Balkema, Rotterdam 1981.

*9365 - SELLS, G.D., KOEPPE, D.E.: Proline oxidation by water-stressed corn shoot
       mitochondria. - Plant Physiol. 65 (Suppl.): 7, 1980.

9366 - SELLS, G.D., KOEPPE, D.E.: Oxidation of proline by mitochondria isolated from
       water-stresses maize shoots. - Plant Physiol. 68: 1058-1063, 1981.

*9367 - SENA GOMES, A.R., KOZLOWSKI, T.T.: Effects of flooding on *Eucalyptus camaldu-
       lensis* and *Eucalyptus globulus* seedlings. - Oecologia 46: 139-142, 1980.

*9368 - SENA GOMES, A.R., KOZLOWSKI, T.T.: Responses of *Pinus halepensis* seedlings to
       flooding. - Can. J. Forest Res. 10: 308-311, 1980.

*9369 - SENA GOMES, A.R., KOZLOWSKI, T.T.: Responses of *Melaleuca quinquenervia* seed-
       lings to flooding. - Physiol. Plant 49: 373-377, 1980.

*9370 - SEYMOUR, V.: Effect of air humidity on cuticular resistance of barberry leaves.
       - Plant Physiol. 65 (Suppl.): 74, 1980.

9371 - SHAH, G.L., ABRAHAM, K.: On the structure and ontogeny of stomata in some
       Umbellifers. - Phyton 21: 189-202, 1981.

9372 - SHARKEY, T.D., RASCHKE, K.: Effect of light quality on stomatal opening in
       leaves of *Xanthium strumarium* L. - Plant Physiol. 68: 1170-1174, 1981.

9373 - SHARKEY, T.D., RASCHKE, K.: Separation and measurement of direct and indirect
       effects of light on stomata. - Plant Physiol. 68: 33-40, 1981.

9374 - SHARMA, G.K., CHANDLER, C., RUPLEY, M., SHAKELFORD, K., PAGE, C.: Geographic
       leaf cuticular and morphological variations in *Trifolium repens* L. (white
       clover). - Fyton 40: 21-26, 1981.

9375 - SHATILOV, I.S., ZAMARAEV, A.G., CHAPOVSKAYA, G.V.: Evapotranspiratsiya i trans-
       piratsiya polevykh kul'tur na dernovo-podzolistoĭ pochve Nechernozem'ya. [E-
       vapotranspiration and transpiration of crops on sod-podzolic soils of Necher-
       nozemye.] Sel'skokhoz. Biol. 16: 387-393, 1981. [In R, ab: E.]

9376 - **SHEEHY, J.E., POPPLE, S.C.:** Photosynthesis, water relations, temperature and canopy structure as factors influencing the growth of sainfoin (*Onobrychis viciifolia* Scop.) and lucerne (*Medicago sativa* L.). - Ann. Bot. 48: 113-128, 1981.

*9377 - **SHEIKH, K.H., HASNAIN, S.:** Iron and manganese relations of flooded and drained plants of *Capsicum annuum* L. - Biologia (Lahore) 24: 387-398, 1978.

9378 - **SHEORAN, I.S., LUTHRA, Y.P., KUHAD, M.S., SINGH, R.:** Effect of water stress on some enzymes of nitrogen metabolism in pigeonpea. - Phytochemistry 20: 2675-2677, 1981.

*9379 - **SHMAT'KO, I.G.:** O roli mineral'nogo pitaniya v optimizatsii vodnogo rezhima rasteniĭ. [Role of mineral nutrition in optimization of water regime of plants.] - In: Mineral'noe Pitanie i Produktivnost' Rasteniĭ. Pp. 15-20. Naukova Dumka, Kiev 1978. [In R.]

*9380 - **SHMAT'KO, I.G., KABLUCHKO, O.I.:** Vplyv obezvodnennya ṇa steblovi merystematychni tkanyny pschenytsi. [Effect of dehydration on the wheat stem meristematic tissues.] - Ukr. bot. Zh. 37 (6): 55-57, 1980. [In Ukr., ab: E, R.]

*9381 - **SHMAT'KO, I.G., SHVEDOVA, O.E., LATASHENKO, O.P., KIRNOS, P.S.:** Posledstvie vliyaniya defitsita vlagi na postuplenie i raspredelenie azota v rasteniyakh ozimoĭ pshenitsy. [After-effect of water deficit on the uptake and distribution of nitrogen in winter wheat plants.] - In: Mineral'noe Pitanie i Produktivnost' Rasteniĭ. Pp. 211-214, 326. Naukova Dumka, Kiev 1978. [In R.]

*9382 - **SIDOROV, V.P.:** Anatomicheskoe stroenie i ust'ichnyĭ apparat lista kormovoĭ svekly. [Anatomical structure and stomatal apparatus of the fodder beet leaf.] - Uch. Zapiski Kazan. gos. pedagog. Inst. 195: 70-75, 1979. [In R.]

*9383 - **SIEGEL, S.M., SIEGEL, B.Z., MASSEY, J., LAHNE, P., CHEN, J.:** Growth of corn in saline waters. - Physiol. Plant. 50: 71-73, 1980.

9384 - **SIEVERDING, E.:** Influence of soil water regimes on VA mycorrhiza. I. Effect on plant growth, water utilization and development of mycorrhiza. - Z. Acker-Pflanzenbau 150: 400-411, 1981.

*9385 - **SIGFRIDSSON, B., ÖQUIST, G.:** Preferential distribution of excitation energy into photosystem I of desiccated samples of the lichen *Cladonia impexa* and the isolated lichen-alga *Trebouxia pyriformis*. - Physiol. Plant. 49: 329-335, 1980.

9386 - **SILEGA, Kh.M., ZAKHARIEV, T.:** Zavisimost voda-dobiv pri razlichna vodoosigurenost na tsarevitsata za z"rno. [The water use efficiency of maize for grain under varying water supply conditions.] - Rasteniev. Nauki 18 (5): 103-111, 1981. [In Bulg, ab: E, R.]

*9387 - **SIMONELLI, M.L.N., SPOMER, L.A.:** Preparation of customized pressure chamber seals for irregularly-shaped, succulent organs. - Agron. J. 72: 699-700, 1980.

*9388 - **SINCLAIR, T.R.:** Plant organ chambers in plant physiology field research. - HortScience 15: 620-622, 1980.

9389 - **SINGER, M.J., MATSUDA, Y., BLACKARD, J.:** Effect of mulch rate on soil loss by raindrop splash. - Soil Sci. Soc. Amer. J. 45: 107-110, 1981.

*9390 - **SINGH, A.I., CHATTERJEE, B.N.:** Barley production under rainfed condition with pre-treated seeds. - Ind. J. Agron. 25: 600-607, 1980.

9391 - **SINGH, A.I., CHATTERJEE, B.N.:** Upland rice production with pre-treated seeds. - Ind. J. agr. Sci. 51: 393-402, 1981.

*9392 - SINGH, G., RAI, V.K.: Responses of two *Cicer arietinum* L. varieties to water stress. - Ind. J. Ecol. 7: 246-253, 1980.

9393 - SINGH, G., RAI, V.K.: Free proline accumulation and drought resistance in *Cicer arietinum* L. - Biol. Plant. 23: 86-90, 1981.

9394 - SINGH, J.P., BHATNAGAR, D.K.: Note on the irrigation of onion as a factor predisposing its bulbs to infection by *Fusarium solani* (Martius) Appel & Wollenweber in storage. - Ind. J. agr. Sci. 51: 686-687, 1981.

*9395 - SINGH, J.S., TRLICA, M.J., RISSER, P.G., REDMANN, R.E., MARSHALL, J.K.: Autotrophic subsystem. - In: BREYMEYER, A.J., VAN DYNE, G.M. (ed.): Grasslands, Systems Analysis and Man. Pp. 59-200. Cambridge University Press, Cambridge, 1980.

9396 - SINGH, K.P., KUMAR, V.: Water use and water-use efficiency of wheat and barley in relation to seeding dates, levels of irrigation and nitrogen fertilization. - Agr. Water Manage. 3: 305-316, 1980/1981.

9397 - SINGH, M.K., SASAHARA, T.: Photosynthesis and transpiration in rice as influenced by soil moisture and air humidity. - Ann. Bot. 48: 513-517, 1981

*9398 - SINGH, S., SHARMA, H.C.: Effect of profile soil moisture and phosphorus levels on the growth yield and nutrient uptake by chickpea. - Ind. J. agr. Sci. 50: 943-947, 1980.

9399 - SINGH, V.P., SINGH, M.: Effect of soil moisture regimes and flag leaf defoliation on yield and yield components of wheat cultivars. - Ind. J. Agron. 26: 7-11, 1981.

*9400 - SINHA, S.K., AGGARWAL, P.K.: Physiological basis of achieving the productivity potential of wheat in India. - Ind. J. Genet. Plant Breed. 40: 375-384, 1980.

*9401 - SINHA, S.K., AGGARWAL, P.K., CHATURVEDI, G.S.: Physiological and biochemical analysis of adaptability in wheat. - In: Proceedings of the Fifth International Wheat Genetics Symposium. Vol. 2. Pp. 946-953. Indian Soc. Genet. Plant Breeding, New Delhi 1978.

9402 - SINHA, S.K., AGGARWAL, P.K., CHATURVEDI, G.S., KOUNDAL, K.R., KHANNA-CHOPRA, R.: A comparison of physiological and yield characters in old and new wheat varieties. - J. agr. Sci. 97: 233-236, 1981.

9403 - SINHA, S.K., NICHOLAS, D.J.D.: Nitrate reductase. - In: PALEG, L.G., ASPINALL, D. (ed.): The Physiology and Biochemistry of Drought Resistance in Plants. Pp. 145-169. Academic Press, Sydney - New York - London - Toronto - San Francisco 1981.

9404 - SIONIT, N., STRAIN, B.R., HELLMERS, H., KRAMER, P.J.: Effects of atmospheric CO_2 concentration and water stress on water relations of wheat. - Bot. Gaz. 142: 191-196, 1981.

9405 - SIPOŞ, G., COIFAN, M., ENCIU, M., ENESCU, D., MORARU, G., PÎNZARIU, D., VASILIU, M.: Reacţia la irigare a porumbului în funcţie de condiţiile ecologice. [Maize reaction to irrigation as influenced by ecological conditions.] - An. Inst. Cercetări Cereale Plante tehnice Fundulea 47: 215-222, 1981. [In Roum, ab: E, R.]

*9406 - SISSON, W.B.: Ambient temperature effects on photosynthesis and root respiration of intact *Yucca elata* Engelm. - Plant Physiol. 65 (Suppl.): 48, 1980.

9407 - SISSON, W.B., BOOTH, J.A., THRONEBERRY, G.O.: Absorption of SO_2 by pecan (*Carya illinoensis* (Wang) K. Koch) and alfalfa (*Medicago sativa* L.) and its effect on net photosynthesis. - J. exp. Bot. 32: 523-534, 1981.

9408 - SIVAKUMAR, M.V.K., SEETHARAMA, N., GILL, K.S., SACHAN, R.C.: Response of sorghum to moisture stress using line source sprinkler irrigation. I. Plant-water
relations. - Agr. Water Manage 3: 279-289, 1981.

9409 - SIVTSEV, M.V., CHECHENINA, T.V., KAZANTSEVA, L.P.: Fiziologicheskie osobennosti
vinograda pri raznykh sposobakh ego orosheniya na krutykh sklonakh. [Physio-
logical peculiarities of grape-vine resulted from different methods of its
irrigation on steep slopes.] - Fiziol. Biokhim. kul't. Rast. 13: 416-421, 1981.
[In R, ab: E.]

*9410 - SKAGGS, R.W.: Combination surface-subsurface drainage systems for humid re-
gions. - J. Irrig. Drain. Div. ASCE 106: 265-283, 1980.

*9411 - SKRE, O., OECHEL, W.C.: Moss production in a black spruce *Picea mariana* forest
with permafrost near Fairbanks, Alaska, as compared with two permafrost-free
stands. - Holarctic Ecol. 2: 249-254, 1979.

9412 - SKRE, O., OECHEL, W.C.: Moss functioning in different taiga ecosystems in inte-
rior Alaska. I. Seasonal, phenotypic, and drought effects on photosynthesis
and response patterns. - Oecologia 48: 50-59, 1981.

9413 - SLAVÍKOVÁ, J.: Differentiation of biomass production on the conic hill Oblík
in the České středohoří mountains. - Preslia 53: 33-44, 1981.

9414 - SLUKHAĬ, S.I., KONONENKO, V.A.: Razvitie i produktivnost' ozimogo yachmenya
pri razlichnom vodoobespechenii. [Development and yield of winter barley under
different water supply.] - Fiziol. Biokhim. kul't. Rast. 13: 594-599, 1981.
[In R, ab: E.]

9415 - SLUKHAĬ, S.I., LATASHENKO, O.P.: Formirovanie belkovogo kompleksa ozimoĭ pshenitsy pri nedostatochnoĭ postoyannoĭ i peremennoĭ vlazhnosti pochvy. [Forma-
tion of winter wheat protein complex with insufficient constant and variable
soil moisture.] - Fiziol. Biokhim. kul't. Rast. 13: 463-470, 1981. [In R,
ab: E.]

*9416 - SLUKHAĬ, S.I., PETRENKO, N.I.: Vodnyĭ rezhim rasteniĭ v svyazi s mineral'nym
pitaniem. [Water regime of plants in connection with mineral nutrition.] -
In: Mineral'noe Pitanie i Produktinost' Rasteniĭ. Pp. 198-206. Naukova Dumka,
Kiev 1978. [In R.]

*9417 - SMAJSTRLA, A.G., HANSON, R.S.: Evaporation effects of sprinkler irrigation
efficiencies. - Soil Crop Sci. Soc. Florida Proc. 39 (2-4): 28-33, 1979.

*9418 - SMITH, J.A.C., MILBURN, J.A.: Solute loading and the control of phloem turgor
in *Ricinus communis* L. - In: SPANSWICK, R.M., LUCAS, W.J., DAINTY, J. (ed.):
Plant Membrane Transport: Current Conceptual Issues. Pp. 543-545. Elsevier/
North-Holland Biomedical Press, Amsterdam - New York - Oxford 1980.

9419 - SMITH, M.K.: Effect of NaCl on the growth of whole plants and their corre-
sponding callus cultures. - Aust. J. Plant Physiol. 8: 267-275, 1981.

9420 - SMITH, R.C.G., HARRIS, H.C.: Environmental resources and restraints to agricultural production in a Mediterranean-type environment. - Plant Soil 58: 31-
57, 1981. Also in: MONTEITH, J., WEBB, C. (ed.): Soil Water and Nitrogen in
Mediterranean-type Environments. Development in Plant and Soil Sciences. Volume 1. Pp. 31-57. Martinus Nijhoff / Dr. W. Junk Publishers, The Hague -
Boston - London 1981.

9421 - SMITH, W.H.: Air Pollution and Forests. Interactions between Air Contaminants
and Forest Ecosystems. (Springer Series on Environmental Management.) -
Springer-Verlag, New York - Heidelberg - Berlin 1981.

9422 - SMITH, W.K.: Temperature and water relation patterns in subalpine understory
plants. - Oecologia 48: 353-359, 1981.

9423 - **SNELGAR, W.P., GREEN, T.G.A.:** Ecologically-linked variation in morphology, acetylene reduction, and water relations in *Pseudocyphellaria dissimilis*. - New Phytol. 87: 403-411, 1981.

9424 - **SNELGAR, W.P., GREEN, T.G.A.:** Carbon dioxide exchange in lichens: Apparent photorespiration and possible role of CO_2 refixation in some members of the *Stictaceae* (Lichenes). - J. exp. Bot. 32: 661-668, 1981.

9425 - **SNELGAR, W.P., GREEN, T.G.A., BELTZ, C.K.:** Carbon dioxide exchange in lichens: Estimation of internal thallus CO_2 transport resistances. - Physiol. Plant. 52: 417-422, 1981.

9426 - **SNELGAR, W.P., GREEN, T.G.A., WILKINS, A.L.:** Carbon dioxide exchange in lichens: Resistances to CO_2 uptake at different thallus water contents. - New Phytol. 88: 353-361, 1981.

9427 - **SOJKA, R.E., STOLZY, L.H.:** Wheat response to drought. - California Agr. 35: 14-15, 1981.

9428 - **SOJKA, R.E., STOLZY, L.H., FISCHER, R.A.:** Seasonal drought response of selected wheat cultivars. - Agron. J. 73: 838-845, 1981.

9429 - **SOLÁROVÁ, J., POSPÍŠILOVÁ, J., SLAVÍK, B.:** Gas exchange regulation by changing of epidermal conductance with antitranspirants. - Photosynthetica 15: 365-400, 1981.

9430 - **SOMMER, C.:** A method for investigating the influence of soil water potential on water consumption, development and yield of plants. - Soil Tilage Res. 1: 163-172, 1980/1981.

9431 - **SOMMER, C., TANASESCU, O.:** Weiterentwicklung der Methode zur kontinuierlichen Wasserversorgung von Versuchsgefässen nach dem Bodenwasserpotential. - Landbauforsch. Völkenrode 31 (1): 59-61, 1981.

*9432 - **SONDGE, V.D., KHUSPE, V.S., ACHARYA, H.S.:** Decision making for irrigation to wheat var. S-308 under certain constraints. - J. Maharasthra agr. Univ. 5: 23-26, 1980.

*9433 - **SONESSON, M. (ed.):** Ecology of a Subarctic Mire. (Ecol. Bull. 30.) Swedish Natural Science Research Council, Stockholm 1980.

9434 - **SOVONICK-DUNFORD, S., LEE, D.R., ZIMMERMANN, M.H.:** Direct and indirect measurements of phloem turgor pressure in white ash. - Plant Physiol. 68: 121-126, 1981.

*9435 - **SPALDING, M.H., EDWARDS, G.E., KU, M.S.B.:** Quantum requirement for photosynthesis in *Sedum praealtum* during two phases of Crassulacean acid metabolism. - Plant Physiol. 66: 463-465, 1980.

*9436 - **SPANSWICK, R.M., LUCAS, W.J., DAINTY, J. (ed.):** Plant Membrane Transport: Current Conceptual Issues. Proceedings of the International Workshop held in Toronto, Canada, July 22-27, 1979. - Elsevier/North-Holland Biomedical Press, Amsterdam - New York - Oxford 1980.

9437 - **SPILKER, D.A., SCHMITTHENNER, A.F., ELLETT, C.W.:** Effects of humidity, temperature, fertility, and cultivar on the reduction of soybean seed quality by *Phomopsis* sp. - Phytopathology 71: 1027-1029, 1981.

*9438 - **SPITTLEHOUSE, D.L., BLACK, T.A.:** Determination of forest evapotranspiration using Bowen ratio and eddy correlation measurements. - J. appl. Meteorol. 18: 647-653, 1979.

*9439 - SPITTLEHOUSE, D.L., BLACK, T.A.: Evaluation of the Bowen ratio/energy balance method for determining forest evapotranspiration. - Atmosphere-Ocean 18: 98-116, 1980.

9440 - SPITTLEHOUSE, D.L., BLACK, T.A.: A growing season water balance model applied to two Douglas fir stands. - Water Resour. Res. 17: 1651-1656, 1981.

9441 - SPRENT, J.I.: Nitrogen fixation. - In: PALEG, L.G., ASPINAL, D. (ed.): The Physiology and Biochemistry of Drought Resistance in Plants. Pp. 131-143. Academic Press, Sydney - New York - London - Toronto - San Francisco 1981.

*9442 - SQUIRE, G.R.: Thermal time and tea. - In: GRACE, J., FORD, E.D., JARVIS, P.G. (ed.): Plants and their Atmospheric Environment. Pp. 363-368. Blackwell Scientific Publications, Oxford - London - Edinburgh - Boston - Melbourne 1980.

9443 - SQUIRE, G.R., BLACK, C.R.: Stomatal behaviour in the field. - In: JARVIS, P.G., MANSFIELD, T.A. (ed.): Stomatal Physiology. Pp. 223-245. Cambridge University Press, Cambridge - London - New York - New Rochelle - Melbourne - Sydney 1981.

9444 - SQUIRE, G.R., BLACK, C.R., GREGORY, P.J.: Physical measurements in crop physiology. II. Water relations. - Exp. Agr. 17: 225-242, 1981.

9445 - STANEV, V., TSONEV, Ts.: Vliyanie na azotniya i fosforniya nedostig v"rkhu difuzionnite s"protivleniya na CO_2, kompensatsionniya punkt i temperaturnata zavisimost na fotosintezata pri sl"nchogleda. [Influence of nitrogen and phosphorus deficit on CO_2 diffusion resistance, compensation point and photosynthetic temperature dependence in sunflower.] - Fiziol. Rast. (Sofia) 5 (3): 12-18, 1979. [In Bulg, ab: E, R.]

9446 - STANHILL, G.: Efficiency of water, solar energy and fossil fuel use in crop production. - In: JOHNSON, C.B. (ed.): Physiological Processes Limiting Plant Productivity. Pp. 39-51. Butterworths, London - Boston - Sydney - Wellington - Durban - Toronto 1981.

9447 - STANLEY, C.D., KASPAR, T.C., TAYLOR, H.M.: Modeling soybean leaf-water potentials under nonlimiting soil water conditions. - Agron. J. 73: 251-254, 1981.

9448 - STANLEY, C.D., ROGERS, J.S., PREVATT, J.W., WATERS, W.E.: Subsurface drainage and irrigation for tomatoes. - Soil Crop Sci. Soc. Florida Proc. 40: 92-95, 1981.

9449 - STAPLES, R.C., ROBINSON, R.W., OEBKER, N.F.: Development of vegetable crops for protected desert environments. - In: MANASSAH, J.T., BRISKEY, E.J. (ed.): Advances in Food-Producing Systems for Arid and Semiarid Lands. Part B. Pp. 737-754. Academic Press, New York 1981.

9450 - STARKEY, T.E., DAVIS, D.D., PELL, E.J., MERRILL, W.: Influence of peroxyacetyl nitrate (PAN) on water stress in bean plants. - HortScience 16: 547-548, 1981.

9451 - STATLER, G.D., NORDGAARD, J.T.: The influence of leaf wetness on wheat leaf rust. - North Dakota Farm Res. 38: 4-5, 1981.

9452 - STAUDINGER, M., ROTT, H.: Evapotranspiration at two mountain sites during the vegetation period. - Nordic Hydrol. 12: 207-216, 1981.

9453 - STEGMAN, E.C., LEMERT, G.W.: Sunflower yield vs. water deficits in major growth periods. - Trans. ASAE 24: 1533-1538, 1545, 1981.

9454 - STEPONKUS, P.L.: Responses to extreme temperatures. Cellular and sub-cellular bases. - In: LANGE, O.L., NOBEL, P.S., OSMOND, C.B., ZIEGLER, H. (ed.): Physiological Plant Ecology I. Responses to the Physical Environment. Pp. 371-402. Springer-Verlag, Berlin - Heidelberg - New York 1981.

*9455 - STEPONKUS, P.L., WIEST, S.C.: Freeze-thaw induced lesion in the plasma mem-
        brane. - In: LYONS, J.M., GRAHAM, D., RAISON, J.K. (ed.): Low Temperature
        Stress in Crop Plants: The Role of Membrane. Pp. 231-254. Academic Press, New
        York - San Francisco - London 1979.

*9456 - STEUDLE, E.: Effect of cell diameter and length on the overall elasticity of
        cylindrical plant cells. Significance for membrane transport and growth. - In:
        SPANSWICK, R.M., LUCAS, W.J., DAINTY, J. (ed.): Plant Membrane Transport: Cur-
        rent Conceptual Issues. Pp. 483-484. Elsevier/North-Holland Biomedical Press,
        Amsterdam - New York - Oxford 1980.

*9457 - STEUDLE, E., FERRIER, J.M., DAINTY, J.: Determination of the transverse exten-
        sion of internodal cell wall tubes of *Chara corallina*. Combination of external
        force method and pressure-probe technique. - In: SPANSWICK, R.M., LUCAS, W.J.,
        DAINTY, J. (ed.): Plant Membrane Transport: Current Conceptual Issues. Pp. 481
        -482. Elsevier/North-Holland Biomedical Press, Amsterdam - New York - Oxford
        1980.

*9458 - STEUDLE, E., SMITH, C., LÜTTGE, U.: Osmotic processes in CAM plants: water-re-
        lation parameters of single leaf cells determined by the pressure-probe tech-
        nique. - In: SPANSWICK, R.M., LUCAS, W.J., DAINTY, J. (ed.): Plant Membrane
        transport: Current Conceptual Issues. Pp. 479-480. Elsevier/North-Holland Bio-
        medical Press, Amsterdam - New York - Oxford 1980.

 9459 - STEUTER, A.A., MOZAFAR, A., GOODIN, J.R.: Water potential of aqueous polyethy-
        lene glycol. - Plant Physiol. 67: 64-67, 1981.

 9460 - STEVENS, M.A.: Resistance to heat stress in crop plants. - In: MANASSAH, J.T.,
        BRISKEY, E.J. (ed.): Advances in Food-Producing Systems for Arid and Semiarid
        Lands. Part A. Pp. 457-486. Academic Press, New York 1981.

*9461 - STEVENSON, T.T., CLELAND, R.E.: Control of osmoregulation in the *Avena* coleo-
        ptile. - Plant Physiol. 65 (Suppl.): 75, 1980.

 9462 - STEVENSON, T.T., CLELAND, R.E.: Osmoregulation in *Avena* coleoptile in relation
        to auxin and growth. - Plant Physiol. 67: 749-753, 1981.

 9463 - STEWART, C.R.: Proline accumulation: Biochemical aspects. - In: PALEG, L.G.,
        ASPINALL, D. (ed.): The Physiology and Biochemistry of Drought Resistance in
        Plants. Pp. 243-259. Academic Press, Sydney - New York - London - Toronto -
        San Francisco 1981.

 9464 - STEWART, M.K., DICKER, M.J.I., JOHNSTON, M.R.: Environmental isotopes in New
        Zealand hydrology 4. Oxygen isotope variations in subsurface waters of the
        Waimea Plains, Nelson. - N. Zeal. J. Sci. 24: 339-348, 1981.

 9465 - STEWART, M.K., TAYLOR, C.B.: Environmental isotopes in New Zealand hydrology 1.
        Introduction: The role of oxygen-18, deuterium, and tritium in hydrology. -
        N. Zeal. J. Sci. 24: 295-311, 1981.

 9466 - STEWART, M.K., WILLIAMS, P.W.: Environmental isotopes in New Zealand hydrology
        3. Isotope hydrology of the Waikoropupu Springs and Takaka River, northwest
        Nelson. - N. Zeal. J. Sci. 24: 323-337, 1981.

*9467 - STOCK, H.-G., EL-NAGGAR, E.-S.: Untersuchungen zur Ermittlung des optimalen
        Beregnungsregimes für Ackerbohnen. - Arch. Acker- Pflanzenbau Bodenk. 24: 665-
        672, 1980.

*9468 - STOCK, H.-G., KAUFHOLD, W., KLEIN, W.: Ergebnisse der Beregnung von Ackerbohnen
        in Feldversuchen auf verschiedenen Standorten der DDR. - Arch. Acker- Pflanzen-
        bau Bodenk. 24: 673-680, 1980.

9469 - STOCK, H.-G., RICHTER, W.: Untersuchungen zur Ermittlung des optimalen Bereg-
nungsregimes für Weisse und Gelbe Lupinen (Körnernutzung). - Arch. Acker-
Pflanzenbau Bodenk. 25: 365-371, 1981.

9470 - STOCK, H.-G., SEIDEL, W., RICHTER, W., STELZNER, C., JAGODA, G.: Ergebnisse
der Beregnung von Weisen und Gelben Lupinen (Körnernutzung) in Feldverssuchen
auf verschiedenen Standorten der DDR. - Arch. Acker- Pflanzenbau Bodenk. 25:
359-364, 1981.

9471 - STOUT, D.G.: Dehydration strain avoidance and tolerance in plant cold hardiness. - J. theor. Biol. 88: 513-521, 1981.

9472 - STOUT, D.G., BROOKE, B., MAJAK, W., REANEY, M.: Influence of cold acclimation
on membrane injury in frozen plant tissue. - Plant Physiol. 68: 248-251, 1981.

*9473 - STROMBERG, E.L., CORDEN, M.E.: Fungitoxicity of xylem extracts from tomato
plants resistant or susceptible to fusarium wilt. - Phytopathology 67: 693-
697, 1977.

*9474 - STROMBERG, E.L., CORDEN, M.E.: Scanning electron microscopy of *Fusarium oxy-
sporum* f. sp. *lycopersici* in xylem vessels of wilt-resistant and susceptible
tomato plants. - Can. J. Bot. 58: 2360-2366, 1980.

*9475 - STUPENDICK, J.T., SHEPHERD, K.R.: Root regeneration of root-pruned *Pinus ra-
diata* seedlings. II. Effects of root-pruning on photosynthesis and translocation. - N. Zeal. J. Forest Sci. 10: 148-158, 1980.

9476 - STUTLER, R.K., JAMES, D.W., FULLERTON, T.M., WELLS, R.F., SHIPE, E.R.: Corn
yield functions of irrigation and nitrogen in Central America. - Irrig. Sci.
2: 79-88, 1981.

*9477 - STYER, R.C., CANTLIFFE, D.J., HALL, C.B.: The relationship of ATP concentra-
tion to germination and seedling vigor or vegetable seeds stored under various
conditions. - J. Amer. Soc. hort. Sci. 105: 298-303, 1980.

9478 - STYLIANOU, Y., ORPHANOS, P.I.: Irrigation of potatoes by sprinklers or trick-
lers on the basis of pan evaporation in a semi-arid region. - Potato Res. 24:
159-170, 1981.

*9479 - SUBRAMANIAM, A.R., PRASADA RAO, G.S.L.H.V.: Meridional variation of water bal-
ance in Rajasthan state. - Ann. Arid Zone 19: 51-57, 1980.

*9480 - SUBRAMANIAM, A.R., PRASADA RAO, G.S.L.H.V.: Climatic study of water balance,
aridity and droughts in Rajasthan state. - Ann. arid Zone 19: 371-377, 1980.

*9481 - SUBRAMANIAM, A.R., SAMBASIVA RAO, A.: Potential evapotranspiration over Maha-
rashtra from climatic data using different methods. - J. Institution Engineers
61: 142-145, 1980.

*9482 - SUDAKOV, V.D., TSEPLYAEVA, M.V., KUZAR, L.L., DYL'KO, A.I., KOROTYSH, V.A.:
Vliyanie urovneĭ pitaniya i vodno-fizicheskikh svoĭstv supeschanoĭ pochvy na
urozhaĭ kartofelya. [Effect of nutrient levels and water and physical properties of a sandy loam soil on potato yields.] - Pochvovedenie 1979 (11): 122-
130, 1979. [In R, ab: E.]

9483 - SUDAR, R.A., SAXTON, K.E., SPOMER, R.G.: A predictive model of water stress in
corn and soybeans. - Trans. ASAE 24: 97-102, 1981.

9484 - SUGAR, D., LOMBARD, P.B.: Pear scab influenced by sprinkler irrigation above
the tree or at ground level. - Plant. Dis. 65: 980, 1981.

9485 - SUHAYDA, C.G., GOODMAN, R.N.: Early proliferation and migration and subsequent xylem occlusion by *Erwinia amylovora* and the fate of its extracellular polysaccharide (EPS) in apple shoots. - Phytopathology 71: 697-707, 1981.

9486 - SUNG, J.-M., COOK, R.J.: Effect of water potential on reproduction and spore germination by *Fusarium roseum* 'Graminearum', 'Culmorum', and 'Avenaceum'. - Phytopathology 71: 499-504, 1981.

9487 - SUR, H.S., PRIHAR, S.S., JALOTA, S.K.: Effect of rice - wheat and maize - wheat rotations on water transmission and wheat root development in a sandy loam of the Punjab, India. - Soil Tillage Res. 1: 361-371, 1980/1981.

*9488 - SUTTON, B.G., DUBBELDE, E.A.: Effects of water deficit on yield of wheat and triticale. - Aust. J. exp. Agr. anim. Husb. 20: 594-598, 1980.

*9489 - SUWANNAPINUNT, W., KOZLOWSKI, T.T.: Effect of SO_2 on transpiration, chlorophyll content, growth, and injury in young seedlings of woody angiosperms. - Can. J. Forest Res. 10: 78-81, 1980.

9490 - SVEINBJÖRNSSON, B., OECHEL, W.C.: Controls on CO_2 exchange in two *Polytrichum* moss species. 1. Field studies on the tundra near Barrow, Alaska. - Oikos 36: 114-128, 1981.

9491 - SWANSON, R.H., WHITFIELD, D.W.A.: A numerical analysis of heat pulse velocity theory and practice. - J. exp. Bot. 32: 221-239, 1981.

9492 - SWINDALE, L.D., BIDINGER, F.R.: Introduction: The human consequences of drought and crop research priorities for their alleviation. - In: PALEG, L.G., ASPINALL, D. (ed.): The Physiology and Biochemistry of Drought Resistance in Plants. Pp. 2-13. Academic Press, Sydney - New York - London - Toronto - San Francisco 1981.

9493 - SWINDALE, L.D., VIRMANI, S.M., SIVAKUMAR, M.V.K.: Climatic variability and crop yields in the semi-arid tropics. - In: BACH, W., PANKRATH, J., SCHNEIDER, S.H. (ed.): Food-Climate Interactions. Pp. 139-166. D. Reidel Publishing Company, Dordrecht - Boston - London 1981.

*9494 - SYVERTSEN, J.P.: Relationships between stomatal conductance and leaf water potential in citrus. - Plant Physiol. 65 (Suppl.): 7, 1980.

9495 - SYVERTSEN, J.P.: Hydraulic conductivity of four commercial citrus rootstock. - J. Amer. Soc. hort. Sci. 106: 378-381, 1981.

*9496 - SYVERTSEN, J.P., ALBRIGO, L.G.: Some effects of grapefruit tree canopy position on microclimate, water relations, fruit yield, and juice quality. - J. Amer. Soc. hort. Sci. 105: 454-459, 1980.

*9497 - SYVERTSEN, J.P., ALBRIGO, L.G.: Seasonal and diurnal citrus leaf and fruit water relations. - Bot. Gaz. 141: 440-446, 1980.

*9498 - SYVERTSEN, J.P., BAUSHER, M.G., ALBRIGO, L.G.: Water relations and related leaf characteristics of healthy and blight affected citrus trees. - J. Amer. Soc. hort. Sci. 105: 431-434, 1980.

9499 - SYVERTSEN, J.P., SMITH, M.L., Jr., ALLEN, J.C.: Growth rate and water relations of citrus leaf flushes. - Ann. Bot. 47: 97-105, 1981.

*9500 - SZAREK, S.R.: Primary production in four North American deserts: indices of efficiency - J. Arid. Environ. 2: 187-209, 1979.

9501 - SZLOVÁK, S., MOLNÁR, Z.: A kukorica levéllemez Fe-, Cu- és Zn-tartalmának
kapcsolata a transzspirációval. [Relationship between the Fe, Cu, Zn content
of maize leaf blades and transpiration.] - Növenytermelés 30: 61-73, 1981 [In:
Hung, ab: E.]

9502 - SZODFRIDT, I.: Further data on the water regime in beech forest types. - Acta
bot. Acad. Sci. Hung. 27: 215-222, 1981.

*9503 - SZUJKÓ-LACZA, J., SEN, S.: Significance of anatomical features of the shoot in
the systematics of Hungarian *Gentiana*. - Acta bot. Acad. Sci. Hung. 25: 365-
402, 1979.

*9504 - TAL, M., KATZ, A.: Salt tolerance in the wild relatives of the cultivated to-
mato: The effect of proline on the growth of callus tissue of *Lycopersicon
esculentum* and *L. peruvianum* under salt and water stresses. - Z. Pflanzenphysiol. 98: 283-288, 1980.

*9505 - TAN, C.S., BLACK, T.A.: Evaluation of a ventilated diffusion porometer for the
measurement of stomatal diffusion resistance of Douglas-fir needles. - Arch.
Meteorol. Geophys. Bioklimatol., Ser. B. 26: 257-273, 1978.

9506 - TAN, C.S., CORNELISSE, A., BUTTERY, B.R.: Transpiration, stomatal conductance,
and photosynthesis of tomato plants with various proportions of root system
supplied with water. - J. Amer. Soc. hort. Sci. 106: 147-151, 1981.

9507 - TAN, C.S., FULTON, J.M.: Estimating evapotranspiration from irrigated crops in
Southwestern Ontario. - Can. J. Plant Sci. 61: 425-435, 1981.

9508 - TAN, C.S., LAYNE, R.E.C.: Application of a simplified evapotranspiration model
for predicting irrigation requirements of peach. - HortScience 16: 172-173,
1981.

*9509 - TANIYAMA, T., NOMURA, T.: [Effects of various synthetic detergents on photo-
synthesis, dry matter and grain production in rice plant.] - Rep. environ. Sci.
Mie Univ. 3: 93-104, 1978. [In Jap, ab: E.]

9510 - TANNER, C.B.: Transpiration efficiency of potato. - Agron. J. 73: 59-64, 1981.

9511 - TANTON, T.W.: Growth and yield of the tea bush. - Exp. Agr. 17: 323-331, 1981.

9512 - TAYLOR, B.H., FERREE, D.C.: The influence of summer pruning on photosynthesis,
transpiration, leaf abscision, and dry weight accumulation of young apple
trees. - J. Amer. Soc. hort. Sci. 106: 389-393, 1981.

9513 - TAYLOR, J.S., REID, D.M., PHARIS, R.P.: Mutual antagonism of sulfur dioxide
and abscisic acid in their effect on stomatal aperture in broad bean (*Vicia
faba* L.) epidermal strips. - Plant Physiol. 68: 1504-1507, 1981.

9514 - TENHUNEN, J.D., LANGE, O.L., BRAUN, M.: Midday stomatal closure in mediterranean type sclerophylls under simulated habitat conditions in an environmental chamber. II. Effect of the complex of leaf temperature and air humidity
on gas exchange of *Arbutus unedo* and *Quercus ilex*. - Oecologia 50: 5-11, 1981.

9515 - TENHUNEN, J.D., LANGE, O.L., PEREIRA, J.S., LÖSCH, R., CATARINO, F.: Midday
stomatal closure in *Arbutus unedo* leaves: Measurements with a steady-state
porometer in the portuguese evergreen scrub. - In: MARGARIS, N.S., MOONEY, H.
A. (ed.): Components of Productivity of Mediterranean-Climate Regions - Basic
and Applied Aspects. Pp. 61-69. Dr. W. Junk Publishers, The Hague - Boston -
London 1981.

*9516 - **TEODORONSKIĬ, V.S., POPOVA, N.Ya.:** Vliyanie antitranspiranta LAT-101 na nekotorye fiziologicheskie protsessy drevesnykh rasteniĭ. [Effect of the antitrans-
pirant LAT-101 on some physiological processes in trees.] - Fiziol Rast. 27:
189-194, 1980. [In R, ab: E.]

*9517 - **TERAMURA, A.H.:** Differences in photosynthetic capacity among three local populations of *Plantago lanceolata* L. - Plant Physiol. 63 (Suppl.): 127, 1979.

9518 - **TERRY, N.:** Physiology of trace element toxicity and its relation to iron
stress. - J. Plant Nutr. 3: 561-578, 1981.

9519 - **TESKEY, R.O., HINCKLEY, T.M.:** Influence of temperature and water potential on
root growth of white oak. - Physiol. Plant. 52: 363-369, 1981.

9520 - **THAKUR, P.S., RAI, V.K.:** Growth characteristics and proline content in relation to water status in two *Zea mays* L. cultivars during rehydration. - Biol.
Plant. 23: 98-103, 1981.

*9521 - **THERIOS, I.N., WEINBAUM, S.A.:** Influence of salinization on growth, mineral
composition, nitrate compensation point and nitrate uptake by two plum clones
grown in solution culture. - Z. Pflanzenphysiol. 99: 305-311, 1980.

9522 - **THIAGARAJAH, M.R., HUNT, L.A., MAHON, J.D.:** Effects of position and age on leaf
photosynthesis in corn (*Zea mays*). - Can. J. Bot. 59: 28-33, 1981.

*9523 - **THIMANN, K.V.:** The senescence of leaves. - In: THIMANN, K.V. (ed.): Senescence
in Plants. Pp. 85-115. CRC Press, Boca Raton 1980.

9524 - **THOMAS, J.E., DAVIDSON, D.:** Effect of ambient water volume on root growth,
cell cycle duration, and mitotic synchrony during germination and seedling
growth of *Vicia faba*. - Can. J. Bot. 59: 1301-1306, 1981.

9525 - **THOMPSON, T.E., FICK, G.W.:** Growth response of alfalfa to duration of soil
flooding and to temperature. - Agron. J. 73: 329-332, 1981.

*9526 - **THORHAUG, A.:** Temperature gradients as a driving force on osmotic flows in a
living membrane. - In: SPANSWICK, R.M., LUCAS, W.J., DAINTY, J. (ed.): Plant
Membrane Transport: Current Conceptual Issues. Pp. 491-492. Elsevier/North-
-Holland Biomedical Press, Amsterdam - New York - Oxford 1980.

9527 - **THORNE, G.N.:** Effects on dry weight and nitrogen content of grains of semi-
-dwarf and tall varieties of winter wheat caused by decreasing the number of
grains per ear. - Ann. appl. Biol. 98: 355-363, 1981.

*9528 - **THORNTON, R.K., WAMPLE, R.L.:** Changes of sunflower in response to water stress
conditions. - Plant Physiol. 65 (Suppl.): 7, 1980.

9529 - **TIBBITS, W.N., BACHELARD, E.P.:** Effects of fertilizer and frequency of watering
on the internal water relations of seedlings of *Angophora costata* and *Banksia
serrata*. - Aust. Forest Res. 11: 23-34, 1981.

9530 - **TIESZEN, L.L., LEWIS, M.C., MILLER, P.C., MAYO, J., CHAPIN, F.S., III. OECHEL,
W.:** An analysis of processes of primary production in tundra growth forms. -
In: BLISS, L.C., CRAGG, J.B., HEAL, D.W., MOORE, J.J. (ed.): Tundra Ecosystems:
A Comparative Analysis. Pp. 285-356. Cambridge University Press, Malta 1981.

9531 - **TIETEMA, T., VAN DER AA, F.:** Ecophysiology of the sand sedge, *Carex arenaria*
L. III. Xylem translocation and the occurrence of patches of vigorous growth
within the continuum of rhizomatous plant system. - Acta bot. Neerl. 30: 183-
189, 1981.

9532 - TIETZ, A., LUDEWIG, M., DINGKUHN, M., DÖRFFLING, K.: Effect of abscisic acid on the transport of assimilates in barley. - Planta 152: 557-561, 1981.

9533 - TILLBERG, E., DONS, C., HAUGSTAD, M., NILSEN, S.: Effect of abscisic acid on CO_2 exchange in *Lemna gibba*. - Physiol. Plant. 52: 401-406, 1981.

9534 - TING, I.P.: Effects of abscisic acid on CAM in *Portulacaria afra*. - Photosynthesis Res. 2: 39-48, 1981.

9535 - TITOV, A.F., TALANOVA, V.V.: Vliyanie khloramfenikola na rost, razvitie i nekotorye fiziologicheskie pokazateli rastenii ogurtsa v rannem ontogeneze. [Effect of chloramphenicol on growth, development and some physiological indexes of cucumber plants during early ontogenesis.] - Ontogenez 12: 503-508, 1981. [In R, ab: E.]

9536 - TKACHUK, E.S., KIRICHENKO, V.P.: Ob ėkonomnom raskhodovanii vody ozimoĭ pshenitseĭ v usloviyakh optimal'nogo termogradienta. [On winter wheat economical water expediture under conditions of optimal thermogradient.] - Fiziol. Biokhim. kul't. Rast. 13: 65-71, 1981. [In R, ab: E.]

*9537 - TOMBESI, L., FAVOLA, G., MORETTI, R., PIRILLO, M., FRANCAVIGLIA, R., COSTANTINI, A.: Studi sulla produttività agricola potenziale e modelli matematici previsionali. [Studies on potential agricultural productivity and previsional mathematical models.] - Ann. Ist. sperim. Nutr. Piante 10 (3): 1-79, 1980. [In Ital. and E.]

9538 - TOMINAGA, Y., TAZAWA, M.: Motive force and rate of cytoplasmic streaming in *Characeae* cells in relation to osmotic pressure and ionic strength of the vacuole and the cytoplasm. - Protoplasma 109: 113-125, 1981.

9539 - TOMOS, A.D., STEUDLE, E., ZIMMERMANN, U., SCHULZE, E.-D.: Water relations of leaf epidermal cells of *Tradescantia virginiana*. - Plant Physiol. 68: 1135-1143, 1981.

*9540 - TONG, Z., LIAN, H.-P., SONG, Y.-R., TAO, G.-Q., CHEN, H.-Y., TSUI, C.: [The relationship between the effects of cytokinin on the expansion and metabolism of excised cucumber cotyledons and water stress.] - Acta bot. Sin. 22: 360-364, 1980. [In Chin, ab: E.]

9541 - TOURAINE, B., GRIGNON, C.: Les fonctions de transport de la racine. II. Les fonctions d'exportation. - Physiol. vég. 19: 581-610, 1981.

9542 - TRACY, T.E., LEWIS, A.J.: Effects of antitranspirants on hydrangea. - HortScience 16: 87-89, 1981.

9543 - TRAVIS, A.J., MANSFIELD, T.A.: Light saturation of stomatal opening on the adaxial and abaxial epidermis of *Commelina communis*. - J. exp. Bot. 32: 1169-1179, 1981.

9544 - TROCHOULIAS, T., MURISON, R.D.: Yield response of bananas to trickle irrigation. - Aust. J. exp. Agr. anim. Husb. 21: 448-452, 1981.

9545 - TROLLDENIER, G.: Influence of soil moisture, soil acidity and nitrogen source on take-all of wheat. - Phytopathol. Z. 102: 163-177, 1981.

9546 - TROLLER, J.A., STINSON, J.V.: Moisture requirements for growth and metabolite production by lactic acid bacteria. - Appl. environ. Microbiol. 42: 682-687, 1981.

*9547 - TROUGHT, M.C.T., DREW, M.C.: The development of waterlogging damage in young wheat plants in anaerobic solution cultures. - J. exp. Bot. 1573-1585, 1980.

*9548 - **TROYER, J.R.**: Diffusion from a circular stoma through a boundary layer. A field-theoretical analysis. - Plant Physiol. 66: 250-253, 1980.

9549 - **TSALOV, Ǐ.**: Vliyanie na faktorite g"stota, torene, khibrid i napoyavane v"rkhu dobiva ot tsarevitsata. [Effect of plant density, fertilizer application, hybrid and irrigation on maize yield.] - Rasteniev. Nauki 18 (5): 53-58, 1981. [In Bulg, ab: E, R.]

9550 - **TSEL'NIKER, Yu.L., MAǏ, V.V., ANDREEVA, T.F.**: Sootnoshenie aktivnosti ribulo-zodifosfatkarboksilazy i intensivnosti fotosinteza u list'ev osiny. [Relationship between ribulose bisphosphate carboxylase activity and photosynthetic rates in aspen leaves.] - Fiziol. Rast. 28: 953-961, 1981. [In R, ab: E.]

9551 - **TSUNO, Y., SUGIMOTO, H.**: [The relation between stomatal aperture with the aid of the infiltration method and photosynthetic rate in some crops.] - Bull. Fac. Agr. Tottori Univ. 33: 126-133, 1981. [In Jap, ab: E.]

9552 - **TU, J.C.**: Effect of salinity on rhizobium-root-hair interaction, nodulation and growth of soybean. - Can. J. Plant Sci. 61: 231-239, 1981.

*9553 - **TULLY, R., MUSGRAVE, M.E., LEOPOLD, A.C.**: Imbibitional chilling damage to soybean and pea: role of the seed coat. - Plant Physiol. 65 (Suppl.): 103, 1980.

9554 - **TULLY, R.E., MUSGRAVE, M.E., LEOPOLD, A.C.**: The seed coat as a control of imbibitional chilling injury. - Crop Sci. 21: 312-317, 1981.

9555 - **TUNSTALL, B.R., CONNOR, D.J.**: A hydrological study of a subtropical semiarid forest of *Acacia harpophylla* F. Muell. ex Benth. (Brigalow). - Aust. J. Bot. 29: 311-320, 1981.

9556 - **TURNER, N.C.**: Techniques and experimental approaches for the measurement of plant water status. - Plant Soil 58: 339-366, 1981.

9557 - **TURNER, N.C.**: Designing crops for dryland Australia: Can the deserts help us? - J. Aust. Inst. agr. Sci. 47: 29-34, 1981.

9558 - **TURNER, N.C., BEGG, J.E.**: Plant-water relations and adaptation to stress. - Plant Soil 58: 97-131, 1981. Also in: MONTEITH, J., WEBB, C. (ed.): Soil Water and Nitrogen in Mediterranean type Environments. Development in Plant and Soil Sciences, Vol. I. Pp. 97-131. Martinus Nijhoff / Dr. W. Junk Publishers, The Hague - Boston - London 1981.

*9559 - **TYERMAN, S.**: Turgor regulation and the development of water potential gradients in *Posidonia*. - In: SPANSWICK, R.M., LUCAS, W.J., DAINTY, J. (ed.): Plant Membrane Transport: Current Conceptual Issues. Pp. 465-466. Elsevier/North-Holland Biomedical Press, Amsterdam - New York - Oxford 1980.

9560 - **TYLER, N.J., FOWLER, D.B., GUSTA, L.V.**: The effect of salt stress on the cold hardiness of winter wheat. - Can. J. Plant Sci. 61: 543-548, 1981.

9561 - **TYLER, N.J., GUSTA, L.V., FOWLER, D.B.**: The effect of a water stress on the cold hardiness of winter wheat. - Can. J. Bot. 59: 1717-1721, 1981.

*9562 - **TYREE, M.T.**: The water permeability of the tonoplast: theoretical aspects. - In: SPANSWICK, R.M., LUCAS, W.J., DAINTY, J. (ed.): Plant Membrane Transport: Current Conceptual Issues. Pp. 459-460. Elsevier/North-Holland Biomedical Press, Amsterdam - New York - Oxford 1980.

9563 - **TYREE, M.T., CRUIZIAT, P., BENIS, M., LoGULLO, M.A., SALLEO, S.**: The kinetics of rehydration of detached sunflower leaves from different initial water deficits. - Plant Cell Environ. 4: 309-317, 1981.

9564 - TYREE, M.T., RICHTER, H.: Alternative methods of analysing water potential isotherms: some cautions and clarifications. I. The impact of non-ideality and of some experimental errors. - J. exp. Bot. 32: 643-653, 1981.

9565 - UNGER, K.: Methoden und Ergebnisse biophysikalisch-ökologischer Untersuchungen pflanzlicher Systeme. - In: UNGER, K., STÖCKER, G. (ed.): Biophysikalische Ökologie und Ökosystemforschung. Pp. 3-23. Akademie-Verlag, Berlin 1981.

9566 - UNGER, K., STÖCKER, G. (ed.): Biophysikalische Ökologie und Ökosystemforschung. - Akademie-Verlag, Berlin 1981.

9567 - UNGER, P.W., JONES, O.R.: Effect of soil water content and a growing season straw mulch on grain sorghum. - Soil Sci. Soc. Amer. J. 45: 129-134, 1981.

*9568 - UNGER, P.W., WIESE, A.F.: Managing irrigated winter wheat residues for water storage and subsequent dryland grain sorghum production. - Soil Sci. Soc. Amer. J. 43: 582-588, 1979.

*9569 - UNSWORTH, M.H.: The exchange of carbon dioxide and air pollutants between vegetation and the atmosphere. - In: GRACE, J., FORD, E.D., JARVIS, P.G. (ed.): Plants and their Atmospheric Environment. Pp. 111-138. Blackwell Scientific Publications, Oxford - London - Edinburgh - Boston - Melbourne 1980.

9570 - UNSWORTH, M.H.: Air pollution and plant productivity. - In: JOHNSON, C.B. (ed.): Physiological Processes Limiting Plant Productivity. Pp. 293-306. Butterworths, London - Boston - Sydney - Wellington - Durban - Toronto 1981.

9571 - UNSWORTH, M.H., BLACK, V.J.: Stomatal responses to pollutants. - In: JARVIS, P.G., MANSFIELD, T.A. (ed.): Stomatal Physiology. Pp. 187-203. Cambridge University Press, Cambridge - London - New York - New Rochelle - Melbourne - Sydney 1981.

9572 - UPPAL, H.S., CHEEMA, S.S.: Effect of mulches and kaolin spray on soil temperature, growth, yield and water use of barley. - Ind. J. agr. Sci. 51: 653-659, 1981.

9573 - URQUHART, A.A., JOY, K.W.: Use of phloem exudate technique in the study of amino acid transport in pea plants. - Plant. Physiol. 68: 750-754, 1981.

*9574 - USHIYAMA, M., CRAVALHO, E.G.: LEVIN, R.L.: Water permeability of yeast cells at sub-zero temperatures. Apendix I. Volumetric changes in yeast cells during freezing at constant cooling rates. - J. Membrane Biol. 46: 112-114, 1979.

9575 - VALIČEK, P.: Some notes on the morphological and anatomical characteristics of the species *Gossypium populifolium* (Benth.) F. Muell ex Tod and *Gossypium pilosum* Fryx (subsection *Grandicalyx* Fryx). - Coton Fibres trop. 36: 233-239, 1981.

*9576 - VAN ALFEN, N.K., ALLARD-TURNER, V.: Susceptibility of plants to vascular disruption by macromolecules. - Plant Physiol. 63: 1072-1075, 1979.

*9577 - VAN ASSCHE, F., CEULEMANS, R., CLIJSTERS, H.: Zinc mediated effects on leaf CO_2 diffusion conductances and net photosynthesis in *Phaseolus vulgaris* L. - Photosynthesis Res. 1: 171-180, 1980.

9578 - VAN AUKEN, O.W., FORD, A.L., ALLEN, J.L.: An ecological comparison of upland deciduous and evergreen forests of central Texas. - Amer. J. Bot. 68: 1249-1256, 1981.

9579 - VAN BAVEL, C.H.M., DAMAGNEZ, J., SADLER, E.J.: The fluid-roof solar greenhouse: energy budget analysis by simulation. Agr. Meteorol. 23: 61-76, 1981.

9580 - **VAN DE DIJK, S.J.:** Two ecologically distinct subspecies of *Hypochaeris radicata* L. III. Differences in drought resistance. - Plant Soil 63: 149-163, 1981.

*9581 - **VAN DER VALK, A.G., DAVIS, C.B.:** The impact of a natural drawdown on the growth of four emergent species in a prairie glacial marsh. - Aquat. Bot. 9: 301-322, 1980.

9582 - **Van KEULEN, H.:** Modelling the interaction of water and nitrogen. - Plant Soil 58: 205-229, 1981. Also in: MONTEITH, J., WEBB, C. (ed.): Soil Water and Nitrogen in Mediterranean-type Environments. Development in Plant and Soil Sciences, Vol. I. Pp. 205-229. Martinus Nijhoff / Dr. W. Junk Publishers, The Hague - Boston - London 1981.

9583 - **Van KEULEN, H., SELIGMAN, N.G., BENJAMIN, R.W.:** Simulation of water use and herbage growth in arid regions - a re-evaluation and further development of the model 'ARID CROP'. - Agr. Systems 6: 159-193, 1980/1981.

*9584 - **VASIĆ, G.:** Uticaj navodnjavanja različitim količinama vode na svojstva klipa i prinos kukuruza na černozemu. [Effect of different amounts of water on ear traits and grain yield of maize on chernozem type of soil.] - Arh. poljoprivredne Nauke 41: 375-384, 1980. [In Croat, ab: E.]

*9585 - **VASIL'EV, B.R., DARIEV, A.S., KOLODYAZHNYĬ, S.F.:** O svyazi mezhdu priznakami stroeniya lista u vidov triby *Hibisceae (Malvaceae)*. [Relations between the leaf characters in the tribe *Hibisceae (Malvaceae)*.] - Bot. Zh. 65: 1188-1195, 1980. [In R, ab: E.]

9586 - **VIDAL, A., ARNAUDO, D., ARNOUX, M.:** La résistance à la sécheresse du soja I. - Influence d'un déficit hydrique sur la croissance et la production. - Agronomie 1: 295-302, 1981.

9587 - **VIDAL, A., ARNOUX, M.:** Drought tolerance processes in soybeans. - Biol. Plant 23: 434-441, 1981.

9588 - **VIGH, L., HORVÁTH, I., FARKAS, T., MUSTÁRDY, L.A., FALUDI-DÁNIEL, Á.:** Stomatal behaviour and cuticular properties of maize leaves of different chilling-resistance during cold treatment. - Physiol. Plant. 51: 287-290, 1981.

9589 - **VITKOV, M.:** Nachini na napoyavane i vodopotreblenie na tsarecitsata pri opodzolen chernozem. [Types of irrigation and water consumption of maize grown on podzolized chernozem soils.] - Rasteniev. Nauki 18 (1): 26-33, 1981. [In Bulg, ab: E, R.]

*9590 - **VLAKHOVA, M., V"LEV, V.:** Prouchvane vliyanieto na napoyavaneto, g"stotata na poseva i khibrida v"rkhu rastezha, razvitieto i dobiva ot tsarevitsata za z"rno v raĭona na yugoiztochna B"lgariya. [A study on the effect of irrigation, planting density and the hybrid on growth, development and yield of grain maize in the south-eastern Bulgaria.] - Rasteniev. Nauki 17 (5): 86-102, 1980. [In Bulg, ab: E, R.]

9591 - **VLEK, P.L.G., FILLERY, I.R.P., BURFORD, J.R.:** Accession, transformation, and loss of nitrogen in soils of the arid region. - Plant Soil 58: 133-175, 1981. Also in: MONTEITH, J., WEBB, C. (ed,): Soil Water and Nitrogen in Mediterranean-type Environments. Development in Plant and Soil Sciences, Vol. I. Pp. 133-175. Martinus Nijhoff / Dr. W. Junk Publishers, The Hague - Boston - London 1981.

9592 - **VON CAEMMERER, S., FARQUHAR, G.D.:** Some relationships between the biochemistry of photosynthesis and the gas exchange of leaves. - Planta 153: 376-387, 1981.

9593 - **VONSHAK, A., RICHMOND, A.:** Photosynthetic and respiratory activity in *Anacystis nidulans* adapted to osmotic stress. - Plant Physiol. 68: 504-505, 1981.

*9594 - **VUČIĆ, N., BOŠNJAK, D.:** Potencijalna evapotranspiracija soje u klimatskim uslovima Vojvodine. [Potential evapotranspiration of soybean grown in climatic conditions of Vojvodina.] - Arh. poljoprivredne Nauke 41: 569-575, 1980. [In Croat, ab: E.]

*9595 - **VUČIĆ, N., DRAGOVIĆ, S., BOŠNJAK, D.:** Produktivnost i kvalitet introdukovanih sorata soje u uslovima navodnjavanja u Vojvodini. [Productivity and quality of introducted soybean varieties under irrigation in Vojvodina.] - Savrem. Poljoprivreda 28: 327-335, 1980. [In Croat, ab: E.]

*9596 - **WAGHMODE, A.P., JOSHI, G.V.:** Glycolate oxidase in halophytes. - Ind. J. exp. Biol. 17: 111-112, 1979.

*9597 - **WAGHMODE, A.P., JOSHI, G.V.:** Kranz leaf anatomy & C_4 dicarboxylic acid pathway of photosynthesis in *Aeluropus lagopoides* L. - Ind. J. exp. Biol. 17: 606-607, 1979.

9598 - **WAGHMODE, A.P., JOSHI, G.V.:** Effect of sodium chloride on photosynthesis in a saline grass *Aeluropus lagopoides* L. - Bot. mar. 24: 361-364, 1981.

9599 - **WAGNER, J., LARCHER, W.:** Dependence of CO_2 gas exchange and acid metabolism of the alpine CAM plant *Sempervivum montanum* on temperature and light. - Oecologia 50: 88-93, 1981.

*9600 - **WAINWRIGHT, S.J.:** Plants in relation to salinity. - Adv. bot. Res. 8: 221-261, 1980.

9601 - **WALGENBACH, R.P., MARTEN, G.C., BLAKE, G.R.:** Release of soluble protein and nitrogen in alfalfa. I. Influence of growth temperature and soil moisture. - Crop Sci. 21: 843-849, 1981.

9602 - **WALKER, M.A., DUMBROFF, E.B.:** Effects of salt stress on abscisic acid and cytokinin levels in tomato. - Z. Pflanzenphysiol. 101: 461-470, 1981.

*9603 - **WALKER, R.R., HAWKER, J.S., TÖRÖKFALVY, E.:** Effect of NaCl on growth, ion composition and ascorbic acid concentrations of capsicum fruit. - Scientia Hort. 12: 211-220, 1980.

9604 - **WALKER, R.R., TÖRÖKFALVY, E., STEELE SCOTT, N., KRIEDEMANN, P.E.:** An analysis of photosynthetic response to salt treatment in *Vitis vinifera*. - Aust. J. Plant. Physiol. 8: 359-374, 1981.

*9605 - **WALLACE, A., ROMNEY, E.M., HUNTER, R.B.:** Carbon fixed in leaves and twigs of field *Larrea tridentata* in two-hour exposure to $^{14}CO_2$. - Great Basin Naturalist Memoirs 1980 (4 - Soil-Plant-Animal Relationships Bearing on Revegetation and Land Reclamation in Nevada Deserts): 121-123, 1980.

9606 - **WALLACE, C.S., SZAREK, S.R.:** Ecophysiological studies of Sonoran desert plants. VII. Photosynthetic gas exchange of winter ephemerals from sun and shade environments. - Oecologia 51: 57-61, 1981.

9607 - **WALLACE, J.S., BATCHELOR, C.H., HODNETT, M.G.:** Crop evaporation and surface conductance calculated using soil moisture data from central India. - Agr. Meteorol. 25: 83-96, 1981.

9608 - **WALSH, J.F., SKUJINS, J.:** Drought effects on the N_2-fixing (acetylene-reducing) ability of vetch and sweetclover growing under saline conditions. - Agron. J. 73: 756-758, 1981.

*9609 - **WALTHER, H.**: *Matudaea menzelii* Walther, ein neues neotropisches Geoelement in der Tertiärflora Mitteleuropas. - Flora 170: 498-516, 1980.

*9610 - **WANG, B.-M., QIN, X., LU, Z.-S.**: [Effect of (2-chloroethyl) trimetylammonium chloride (CCC) on water status of wheat under different water regimes.] - Acta bot. Sin. 22: 297-299, 1980. [In Chin.]

*9611 - **WANG, W.-Y.**: [Observations on the capacity of drought resistance of *Sophora moocroftiana* (Wall.) Benth. ex Baker.] - Acta bot. Sin. 22: 293-294, 1980. [In Chin.]

9612 - **WARDLE, K., SHORT, K.C.**: Induced stomatal responses in epidermal strips of *Vicia faba* L. - J. biol. Educ. 15: 117-122, 1981.

9613 - **WARDLE, K., SHORT, K.C.**: Responses of stomata in epidermal strips of *Vicia faba* to carbon dioxide and growth hormones when incubated on potassium chloride and potassium iminodiacetate. - J. exp. Bot. 32: 303-309, 1981.

9614 - **WARDLE, P.**: Winter desiccation of conifer needles simulated by artificial freezing. -Arctic Alpine Res. 13: 419-423, 1981.

9615 - **WAREING, P.F., PHILLIPS, I.D.J.**: Growth and Differentiation in Plants. Third Edition. Pergamon Press, Oxford - New York - Toronto - Sydney - Paris - Frankfurt 1981.

*9616 - **WATERKEYN, L., BIENFAIT, A.**: Production et dégradation de callose dans les stomates des fougères. - Cellule 73: 83-97, 1979.

9617 - **WATSON, B.T., WARDLAW, I.F.**: Metabolism and export of ^{14}C-labelled photosynthate from water-stressed leaves. - Aust. J. Plant. Physiol. 8: 143-153, 1981.

9618 - **WATTS, S., RODRIGUEZ, J.L., EVANS, S.E., DAVIES, W.J.**: Root and shoot growth of plants treated with abscisic acid. - Ann. Bot. 47: 595-602, 1981.

9619 - **WEISS, A., LUKENS, D.L.**: Electronic circuit for detecting leaf wetness and comparison of two sensors. - Plant Dis. 65: 41-43, 1981.

9620 - **WEISSHAUPT, F., FRIELINGHAUS, M.**: Zusammenhang zwischen Regenintensität und Ertrag am Beispiel der Kreisberegnungsmaschine "Fregat". - Arch. Acker-Pflanzenbau Bodenk. 25: 125-133, 1981.

9621 - **WENDA, W.I., HANKS, R.J.**: Corn yield and evapotranspiration under simulated drought conditions. - Irrig. Sci. 2: 193-204, 1981.

9622 - **WENKERT, W.**: The behavior of osmotic potential in leaves of maize. - Environ. exp. Bot. 21: 231-239, 1981.

9623 - **WENKERT, W., FAUSEY, N.R., WATTERS, H.D.**: Flooding response in *Zea mays* L. - Plant Soil 62: 351-366, 1981.

9624 - **WERGER, M.J.A., ELLIS, R.P.**: Photosynthetic pathways in the arid regions of South Africa. - Flora 171: 64-75, 1981.

*9625 - **WEST, D.W., GAFF, D.F.**: Changes in diffusion resistance of apple leaves undergoing rapid fluctuations in leaf temperature. - Physiol. Plant. 48: 578-583, 1980.

9626 - **WEYERS, J.D.B., TRAVIS, A.J.**: Selection and preparation of leaf epidermis for experiments on stomatal physiology. - J. exp. Bot. 32: 837-850, 1981.

9627 - **WHITEHEAD, D., JARVIS, P.G.:** Coniferous forests and plantations. - In: KOZ-LOWSKI, T.T. (ed.): Water Deficits and Plant Growth. Vol. VI. Woody Plant Communities. Pp. 49-152. Academic Press, New York - San Francisco - London 1981.

9628 - **WHITEHEAD, D., OKALI, D.U.U., FASEHUN, F.E.:** Stomatal response to environmental variables in two tropical forest species during the dry season in Nigeria. - J. appl. Ecol. 18: 571-587, 1981.

* 9629 - **WHITING, B.H., VAN DE VENTER, H.A., SMALL, G.C.:** Crassulacean acid metabolism in jointed cactus (*Opuntia aurantiaca* Lindley). - Agroplantae 11 (2): 41-43, 1979.

*9630 - **WHITLOCK, A.J.:** The effect of anti-transpirants and post planting watering on the establishment and yield of celery transplants. - Exp. Hort. 31: 21-25, 1979.

9631 - **WIEBE, H.H.:** Measuring water potential (activity) from free water to oven dryness. - Plant Physiol. 68: 1218-1221, 1981.

9632 - **WIEBOLD, W.J., SHIBLES, R., GREEN, D.E.:** Selection for apparent photosynthesis and related leaf traits in early generations of soybeans. - Crop Sci. 21: 969-973, 1981.

9633 - **WIENCKE, C., LÄUCHLI, A.:** Inorganic ions and floridoside as osmotic solutes in *Porphyra umbilicalis*. - Z. Pflanzenphysiol. 103: 247-258, 1981.

9634 - **WIGHT, J.R., HANKS, R.J.:** A water balance, climate model for range herbage production. - J. Range Manage. 34: 307-311, 1981.

9635 - **WIGNARAJAH, K., BAKER, N.R.:** Salt induced responses of chloroplast acivities in species of differing salt tolerance. Photosynthetic electron transpor in *Aster tripolium* and *Pisum sativum*. - Physiol. Plant. 51: 387-393, 1981.

9636 - **WILCOX, D.A., DAVIES, F.S.:** Temperature-dependent and diurnal root conductivities in two citrus rootstocks. - HortScience 16: 303-305, 1981.

9637 - **WILCOX, J.R., SEDIYAMA, T.:** Interrelationships among height, lodging and yield in determinate and indeterminate soybeans. - Euphytica 30: 323-326, 1981.

9638 - **WILD, A., FORSCHNER, W., ZERBE, R., RÜHLE, W.:** The effect of kinetin on the transpiration and the photosynthetic capacity of primary leaves of *Sinapis alba*. - Z. Pflanzenphysiol. 105: 93-96, 1981.

9639 - **WILLIAMS, G.M., AYRES, P.G.:** Effects of powdery mildew and water stress on CO_2 exchange in uninfected leaves of barley. - Plant Physiol. 68: 527-530, 1981.

9640 - **WILLMER, C.M.:** Guard cell metabolism. - In: JARVIS, P.G., MANSFIELD, T.A. (ed.): Stomatal Physiology. Pp. 87-102. Cambridge University Press, Cambridge - London - New York - New Rochelle - Melbourne - Sydney 1981.

9641 - **WILSON, J.A.:** Stomatal responses to applied ABA and CO_2 in epidermis detached from well-watered and water-stressed plants of *Commelina communis* L. - J. exp. Bot. 32: 261-269, 1981.

9642 - **WINTER, K.:** CO_2 and water vapour exchange, malate content and $\delta^{13}C$ value in *Cicer arietinum* grown under two water regimes. - Z. Pflanzenphysiol. 101: 421-430, 1981.

9643 - **WOODSTOCK, L.W., TAO, K.-L.J.:** Prevention of imbibitional injury in low vigor soybean embryonic axes by osmotic control of water uptake. - Physiol. Plant. 51: 133-139, 1981.

*9644 - **WOODWARD, F.I.**: Shoot extension and water relations of *Circaea lutetiana* in sunflecks. - In: GRACE, J., FORD, E.D., JARVIS, P.G. (ed.): Plants and their Atmospheric Environment. Pp. 83-91. Blackwell Scientific Publications, Oxford - London - Edinburgh - Boston - Melbourne 1980.

9645 - **WRIGHT, S.T.C.**: The effect of light and dark periods on the production of ethylene from water-stressed wheat leaves. - Planta 153: 172-180, 1981.

9646 - **WYN JONES, R.G.**: Salt tolerance. - In: JOHNSON, C.B. (ed.): Physiological Processes Limiting Plant Productivity. Pp. 271-292. Butterworths, London - Boston - Sydney - Wellington - Durban - Toronto 1981.

9647 - **WYN JONES, R.G., STOREY, R.**: Betaines. - In: PALEG, L.G., ASPINALL, D. (ed.): The Physiology and Biochemistry of Drought Resistance in Plants. Pp. 171-204. Academic Press, Sydney - New York - London - Toronto - San Francisco 1981.

9648 - **XU, X.-D., ZHU, H.-S.**: [Effects of Tween-80 treatment on stomatal movement and transpiration of maize seedlings.] - Acta phytophysiol. Sin. 7: 121-127, 1981. [In Chin, ab: E.]

9649 - **YABLONSKIĬ, E.A.**: Dinamika soderzhaniya vody v generativnykh pochkakh i odno-letnikh pobegakh razlichnykh po zimostoĭkosti sortov persika. [Dynamics of water content in generative buds and annual sprouts of peach varieties differing in frost resistance.] - Fiziol. Biokhim. kul't. Rast. 13: 200-205, 1981. [In R, ab: E.]

9650 - **YAGUCHI, Y., KONDO, K.**: Stomatal responses to prey capture and trap narrowing in Venus's flytrap (*Dionaea muscipula* Ellis). - Fyton 41: 83-90, 1981.

9651 - **YAKLICH, R.W., CREGAN, P.B.**: Moisture migration into soybean pods. - Crop Sci. 21: 791-793, 1981.

9652 - **YAMAGISHI, A., SATOH, K., KATOH, S.**: The concentrations and thermodynamic activities of cations in intact *Bryopsis* chloroplasts. - Biochim. biophys. Acta 637: 252-263, 1981.

9653 - **YAMAGUCHI, H., WATANABE, M., SATO, S., KANBAYASHI, Y.**: Yield response of an erect, narrow and thick-leaved rice mutant and the derived strain to different planting densities. - In: Induced Mutations - a Tool in Plant Research. Pp. 201 201-211. International Atomic Energy Agency, Vienna 1981.

9654 - **YAMAMOTO, T., WATANABE, S., ABE, Y.**: [Water balance in pear trees.] - J. Jap Soc. hort. Sci. 50: 297-305, 1981. [In Jap, ab: E.]

*9655 - **YARISH, C., EDWARDS, P., CASEY, S.**: The effects of salinity, and calcium and potassium variations on the growth of two estuarine red algae. - J. exp. Mar. Biol. Ecol. 47: 235-249, 1980.

9656 - **YATES, D.J.**: Effect of the angle of incidence of light on the net photosynthesis rates of *Sorghum almum* leaves. - Aust. J. Plant Physiol. 8: 335-346, 1981.

*9657 - **YEATON, R.I., YEATON, R.W., HORENSTEIN, J.E.**: The altitudinal replacement of Digger pine by Ponderosa pine on the western slopes of the Sierra Nevada. - Bull. Torrey bot. Club 107: 487-495, 1980.

*9658 - **YELENOSKY, G.**: Effect of moisture on leaves and wood of young citrus trees during controlled freezes. - Proc. Florida State hort. Soc. 93: 3-5, 1980.

*9659 - **YODER, P.G., LABER, L.J.**: After-effects of water stress on potatoes (*Solanum tuberosum* L.) - Plant Physiol. 65 (Suppl.): 9, 1980.

*9660 - YOUNG, D.R., SMITH, W.K.: Influence of sunlight on photosynthesis, water rela-
tions, and leaf structure in the understory species *Arnica cordifolia*. -
Ecology 61: 1380-1390, 1980.

9661 - YOUNG, E., HAND, J.M., WIEST, S.C.: Diurnal variation in water potential components and stomatal resistance of irrigated peach seedlings. - J. Amer. Soc.
hort. Sci. 106: 337-340, 1981.

9662 - YOUNG, J.E.: The use of canonical correlation analysis in the investigation of
relationships between plant growth and environmental factors. - Ann. Bot. 48:
811-825, 1981.

*9663 - YOUNG, R.H.: Water movement in limbs, trunks and roots of healthy and blight-
-affected cultivar 'Valencia' orange trees. - Proc. Florida State hort. Soc.
92: 64-67, 1979.

9664 - YURANICH, N., MATSURA, S., SREICH, R., DZHORDZHEVICH, L., VUCHELICH, D.:
Issledovanie sostoyaniya vody v kletkakh *Nitella mucronata* metodom yadernogo
magnitnogo rezonansa. [Proton spin relaxation of water in *Nitella mucronata*
cells.] - Biofizika 25: 1011-1016, 1980. [In R, ab: E.]

9665 - ZABLOTOWICZ, R.M., FOCHT, D.D., CANNELL, G.H.: Nodulation and N-fixation of
field-grown california cowpeas as influenced by well-irrigated and droughted
conditions. - Agron. J. 73: 9-12, 1981.

9666 - ZAVITKOVSKI, J., JEFFERS, R.M., NIENSTAEDT, H., STRONG, T.F.: Biomass production of several jack pine provenances at three Lake States locations. - Can. J.
Forest Res. 11: 441-447, 1981.

9667 - ZEIGER, E.: Novel approaches to the biology of stomatal guard cells: proto-
plast and fluorescence studies. - In: JARVIS, P.G., MANSFIELD, T.A. (ed.):
Stomatal Physiology. Pp. 103-117. Cambridge University Press, Cambridge -
London - New York - New Rochelle - Melbourne - Sydney 1981.

9668 - ZEIGER, E., ARMOND, P., MELIS, A.: Fluorescence properties of guard cell
chloroplasts. Evidence for linear electron transport and light-harvesting
pigments of photosystems I and II. - Plant Physiol. 67: 17-20, 1981.

9669 - ZENIŠČEVA, L.: Odrůdově specifická reakce jarního ječmene na vysoké dávky du-
síku, fosforu a draslíku při diferencovaném vodním režimu půdy. [The varie-
tally specific response of spring barley to high application rates of nitrogen, phosphorus and potassium in soils with different moisture regimes.] -
Rost. Výroba (Praha) 27: 683-692, 1981. [In Czech, ab: E, G, R.]

9670 - ZHOLKEVICH, V.N., SINITSINA, Z.A., PEISAKHZON, B.I.: On physiological regulation of water transport in root systems. - Stud. biophys. 85: 17-18, 1981.

*9671 - ŽÍDEK, V., ČERMÁK, J., ÚLEHLA, J.: Transpirační proud v dubu v lužním lese ve
vztahu k potenciální evapotranspiraci. [Relation of transpiration flow rate
in oak tree and potential evapotranspiration in the flood-plain forest.] - In:
Zborník Referátov 3. Zjazdu Slovenskej Botanickej Spoločnosti Zvolen 1980. Pp.
227-232. Slovenská botanická spoločnosť, Bratislava 1980. [In Czech, ab: E.]

9672 - ZIEGLER, H.: Einige differenzierte Zelltypen im Pflanzenreich. - In: METZNER,
H. (ed.): Die Zelle. Struktur und Funktion. Pp. 90-120. Wissenschaftliche
Verlagsgesellschaft mbH, Stuttgart 1981.

9673 - ZIEGLER, H., LIN, H.-P.P., SCHNABL, H.: Änderungen in der intra- und inter-
cellulären Kompartimentierung. - Ber. Deut. bot. Ges. 94: 193-202, 1981.

9674 - **ZILINSKAS, B.A., GLICK, R.E.**: Noncovalent intermolecular forces in phycobili-
somes of *Porphyridium cruentum*. - Plant Physiol. 68: 447-452, 1981.

*9675 - **ZIMMERMANN, U., HÜSKEN, D.**: Elastic properties of the cell wall of *Halicystis
parvula*. - In: SPANSWICK, R.M., LUCAS, W.J., DAINTY, J. (ed.): Plant Membrane
Transport: Current Conceptual Issues. Pp. 471-472. Elsevier/North-Holland
Biomedical Press, Amsterdam - New York - Oxford 1980.

9676 - **ZYALALOV, A.A.**: Polyarnost' vodnoĭ provodimosti épidermisa. [Polar permeability of epidermis to water.] - Fiziol. Rast. 28: 982-986, 1981. [In R, ab: E.]

AUTHORS' INDEX

Authors' names are presented in the form in which they appear in the respective
publication. The names from papers published in Cyrillic character are transcribed
as shown in Instructions for Use. Alternative spelling and form of the name of the
same author are usually cross-indexed.

ASPINALL, D. 8182, 9134, 9316
ASSCHE, F., van see Van ASSCHE, F.
ATANASIU, N. 8183
ATKINS, C.A. 8682, 8904
ATSMON, D. 8973
AUCLAIR, D. 8184
AUKEN, O.W., van see VAN AUKEN, O.W.
AUST, H.-J. 8185
AUSTIN, R.B. 8186
AVITA, S. 8187
AXELSSON, B. 8143
AYRES, P.G. 8188, 8189, 8190,
9639

B

BAAS, P. 8191, 9100
BABALOLA, O. 8192
BACA, F. 8193
BACH, T. 8915
BACHELARD, E.P. 9529
BAČIĆ, G. 9248
BADANOVA, K.A. 8194
BAEUMER, K. 8472
BAHN, E. 8305
BAIER, W. 8195
BAILEY, H.P. 8196
BAILEY, W.G. 8197, 8198
BAJAJ, K.L. 8430
BAKER, E.A. 8199
BAKER, G.A. 9149
BAKER, N.R. 9016, 9635
BAKRADZE, N.G. 8200, 8201
BALASUBRAMANIAN, V. 8202
BALDOCCHI, D.D. 8203, 8204, 8205
BALINA, N.V. 8194
BALL, E. 9286
BALLA, Yu.I. 8200
BALLARD, T.M. 8443
BAMATRAF, A.M. 8318
BANBA, H. 8206, 8207
BARAK, P. 8343
BARANOVA, M.A. 8208
BARAVIKOVA, A.M. 8209
BARBER, J. 8210
BARFIELD, B.J. 9046, 9047
BARKER, A.V. 8211
BARKER, N.A. 8663
BARKER, W.G. 9168
BARLOW, E.W.R. 9173
BARNI, N.A. 9198, 9199
BARRAN, L.R. 8212
BARRENTINE, W.L. 8631
BARRETT, J.E. 8760
BARRICK, W.E. 8213
BARRS, H.D. 8860
BARTA, A.L. 8214
BARTHLOTT, W. 8215

BARTOLOMAEUS, W. 8655
BARTON, P.G. 8462
BARUCH, Z. 8216
BAR-YOSEF, B. 8779
BASHA, S.K.M. 9241
BATAL, K.M. 8217
BATANOUNY, K.H. 8218, 8219, 8220
BATCHELOR, C.H. 9607
BATES, L.M. 8221
BÄTZ, G. 8305
BAUDER, J.W. 8222
BAUER, H. 8895
BAUMEISTER, W. 8223
BAUSHER, M.G. 9498
BAVEL, C.H.M., van see VAN BAVEL, C.H.M.
BAZZAZ, F.A. 8981
BEADLE, C.L. 8224
BEALL, P.T. 8225
BECWAR, M.R. 8226
BEDELL, T.E. 8552
BEDENKO, V.P. 8227
BEDUNAH, D. 8228
BEGG, J.E. 9558
BEHRENDT, S. 8842
BELEITES, F. 8512
BELFORD, R.K. 8229
BELIĆ, J. 8230
BELL, C.J. 8231, 8232, 8233
BELL, D.T. 8579
BELL, K.R. 8234
BELOKOBYL'SKII, I.M. 8235
BELTZ, C.K. 9425
BENECKE, P. 8236, 8237
BENECKE, U. 8238
BENGTSON, C. 8239
BENIS, M. 9563
BENJAMIN, R.W. 9583
BENKENSTEIN, H. 8864
BENLLOCH, M. 8425
BENNETT, J.H. 8240, 8656
BENNETT, J.M. 8153, 8241, 8242
BENOIT, G. 8581
BENTRUP, F.-W. 8302
BENZ, L.C. 9263
BÉRAUD, J. 9162
BERGAMASCHI, H. 9198, 9199
BERINGER, H. 8624
BERNIER, G. 8243
BERNSTEIN, C.S. 9065
BERRY, J. 8270
BERSHTEĬN, B.I. 8244, 8719
BEWLEY, J.D. 8245, 8246, 8274
9353, 9354
BHARDWAJ, R. 8247
BHARTI, S. 9205
BHATNAGAR, D.K. 9394
BHATT, D.C. 8775
BHATTACHARYYA, A.C. 9138
BIDINGER, F. 8512

KAYE, P.E. 8937
KAZANTSEVA, L.P. 9409
K"DREV, T.G. 8805
KEEFE, P.D. 8806
KEIM, D.L. 8807
KEITH, A.D. 8808
KELLEY, D.B. 8491
KELLIHER, F.M. 8359, 8809
KEMP, P.R. 8810, 8811
KENASCHUK, E.O. 8609
KENG, J.C.W. 8812
KENNEDY, R.A. 8813
KENYON, W.H. 8814
KESHELASHVILI, L.V. 8201
KESSLY, D.S. 8815
KEULEN, H., van see Van KEULEN, H.
KEYS, C.H. 8828
KHALIFA, M.A. 8129
KHAN, A.A. 8816
KHAN"MOVA, T. 8817
KHANNA-CHOPRA, R. 8818, 9402
KHARANYAN, N.N. 9171
KHARE, P.K. 8819
KHARLAMOVA, N.V. 8255
KHERA, K.L. 9208
KHODARY, S.E.A. 9322
KHOSLA, B.K. 8472
KHOTYLEVA, L.V. 8463
KHRISTOV, I. 8856
KHUSPE, V.S. 9432
KHVOSTOVA, I.V. 8820
KIM, J.H. 8821
KIMES, D.S. 8822
KIMMERER, T.W. 8823
KIMURA, M. 8729
KINCAID, D.T. 8824
KINET, J.-M. 8243
KINGSBURY, R.W. 8491
KINGSOLVER, J.G. 8825
KIRICHENKO, V.P. 9536
KIRILLINA, V.I. 9171
KIRKHAM, M.B. 8137, 8826, 8827
KIRKLAND, K.J. 8828
KIRNOS, P.S. 9381
KIRST, G.O. 8257, 8258, 8259,
8829, 8830
KISHITANI, S. 8831
KISLYUK, I.M. 8832
KLAR, A.E. 9283
KLEIN, A.O. 8833
KLEIN, M. 8834
KLEIN, S. 8156
KLEIN, W. 9468
KLEINKOPF, G.E. 8352, 8460, 8461
KLEINOVÁ, M. 8401
KLOCKARE, R. 8835
KLOSSON, R.J. 8836
KLUGE, M. 8837, 8838
KLUTHCOUSKI, J. 8408
KLYSHEV, L.K. 8157

KNAPP, A.K. 8839
KNECHT, G.N. 9106
KNIGHT, D.H. 8840, 8841
KNITTEL, H. 8842
KOBATA, T. 8843
KOBAYASHI, K. 8844
KOBAYASHI, Y. 9018, 9019
KOCH, E.J. 8656
KOEPPE, D.E. 9365, 9366
KOLARI, K. 8708
KOLBASINA, E.I. 8845
KOLCHINA, N.A. 8520
KOLESNIK, T.I. 9171
KOLEV, B. 8856
KOLODYAZHNYĬ, S.F. 9585
KOMATSU, H. 9069
KOMPANIETS', I.I. 9309
KONDO, K. 9650
KONDO, N. 8846
KONONENKO, V.A. 9414
KÖPKE, U. 8472
KORBAN, S.S. 8847
KORMAKOVSKIĬ, A.Ya. 8848
KÖRNER, C. 8849, 8850, 8851
KOROTYSH, V.A. 9482
KORZINNIKOV, Yu.S. 8547
KOUCHKOVSKY, Y., de 8852
KOUNDAL, K.R. 9402
KOVACHEVA, I. 8853
KOZEL, U. 8915
KOZLOWSKI, T.T. 8823, 8854, 8855,
8857, 9087, 9140, 9141, 9367,
9368, 9369, 9489
KRAFTI, G. 8856
KRAMER, D. 8436
KRAMER, P.J. 8857, 9404
KRANTZ, B.A. 8858
KRANZ, J. 8459
KRAUSE, C.R. 8859
KRAUSE, G.H. 8836, 9332
KRIEDEMANN, P.E. 8860, 8861, 9604
KROGMAN, K.K. 8681
KROLL, R.G. 8862
KRONSTAD, W.E. 8807
KRSTIĆ, B. 8863
KRÜGER, W. 8864
KRUPYANSKIĬ, Yu.F. 8865
KRYMSKAYA, N.B. 8547
KRYUKOVA, E.V. 8874
KU, M.S.B. 9435
KUBICHEK, S.A. 8866
KUCERA, J. 8330, 8331
KUEH, J.S.H. 8867
KUHAD, M.S. 9378
KUIPER, F. 8868
KUIPER, P.J.C. 8957, 9187
KULL, U. 8688
KUMAR, D. 8869
KUMAR, V. 9396
KUPKANCHANAKUL, T. 8870

PLANT INDEX

This index contains plant genera and types interesting as experimental material
for physiological, ecological and agricultural studies. The Latin plant names are
the main items which present the reference number. English names of the most common
plants are cross-indexed.

A

Abelmoschus 9252

Abies 8226, 8253, 8365, 8742,
8743, 8762, 8800, 8801, 8802,
8839, 8857, 8894, 9132, 9220,
9229, 9358, 9422, 9627

Abrus 9318

Abutilon 8360

Acacia 8579, 8772, 8857, 8861,
8939, 8985, 9183, 9555, 9646

Acalypha 9243

Acer 8213, 8366, 8433, 8643,
8677, 8709, 8854, 8857, 8860,
8861, 8929, 8933, 9489, 9516

Achillea 9094

Achyranthes 9243

Aconitum 8187

Actaea 8187

Actinidia 8504, 8505

Adenocarpus 9318

Aegiceras 9600

Aeluropus 9596, 9597, 9598

Aeschynomene 8153, 9008

Aesculus 8517, 8915, 9578

Agave 8861, 9085, 9349

Agropyron 8311, 8531, 8552,
8607, 8644, 8796, 8984, 9094,
9204, 9252, 9600, 9646

Agrostis 8655, 9252, 9600

Alchemilla 8850

alder see *Alnus*

alfalfa see *Medicago*

Algae
    Anabaena 8162, 8452, 8453,
    8492, 9234
    Anacystis 8492, 9593
    Bostrychia 8415, 9655
    Bryopsis 8256, 9652
    Ceramium 8256
    Chaetomorpha 8256
    Chara 8155, 8749, 8932,
    9330, 9331, 9457, 9538
    Chlamydomonas 8492
    Chlorella 8281, 8338, 8492,
    9463
    Codium 8256
    Corallina 8922
    Cyclotella 8281, 9463
    Dunaliella 8281, 8492, 8573,
    8574, 8815
    Enteromorpha 9310
    Fucus 8281, 8683, 8771
    Gelidium 8922
    Gigartina 8922
    Gracilaria 8492, 8830
    Halicystis 8256, 9675
    Isochrysis 8884
    Laminaria 8771
    Lamprothamnium 8257, 8258, 8259
    Macrocystis 8922, 9346
    Nitella 8256, 8377, 9248, 9562
    9664
    Nostoc 8452, 8492

Algae (continued)
   Ochromonas 8281
   Oscillatoria 8492
   Pelvetia 8158
   Phaeodactylum 8281, 9350, 9351
   Platymonas 8281
   Porphyra 9256, 9257, 9258, 9259, 9260, 9633
   Porphyridium 8281, 9674
   Rhodymenia 8830
   Scenedesmus 9385
   Stichococcus 8281
   Synechococcus 8282, 8492
   Trebouxia 9385
   Ulva 8434, 8492, 8683
   Valonia 8256, 8612, 8749

Alisma 8947, 9049

Allium 8158, 8182, 8303, 8405, 8442, 9098, 9101, 9135, 9136, 9142, 9184, 9252, 9282, 9312, 9343, 9349, 9356, 9394, 9477, 9626, 9640, 9667, 9668, 9676

Alnus 8677, 8722, 8762, 8850, 8857, 8861, 8933

almond see *Amygdalus*

Aloë 8397

Alopecurus 9252

Alstonia 9159

Alternanthera 9156, 9243

Amaranthus 8307, 8561, 8686, 8745, 8857, 9131, 9160, 9243, 9341, 9429, 9617, 9642

Ambrosia 8352, 8939, 8985

Amelanchier 9472

Ammi 9371

Amygdalus 8175, 8648

Anacharis 9523

Anagyris 9318

Ananas 9090

Anastatica 8245, 8535

Andromeda 9433

Andropogon 9006

Anemone 8187, 8475

Anethum 9371

Anthoceros 8929

Anthyllis 8850, 9318

Apium 8164, 8248, 9630

apple see *Malus*

apricot see *Armeniaca*

Aquilegia 8187, 9094

Arachis 8202, 8241, 8245, 8342, 8370, 8846, 9088, 9239, 9241, 9252, 9317, 9326, 9429, 9443, 9493

arborvitae see *Thuja*

Arbutus 9514, 9515

Arctostaphylos 8365, 8735, 9011, 9099, 9149, 9194, 9195

Argemone 8620

Aristida 8493

Armeniaca 8646, 8849, 8951

Arnica 8850, 9094, 9422, 9660

Arrhenatherum 8435, 8850

Artemisia 8182, 8352, 8636, 8939, 9072, 9094, 9195, 9360

Arum 8475

Asarum 8877

Asclepias 9120

ash see *Fraxinus*

Asparagus 9098

aspen see *Populus*

Aster 8588, 8850, 9635, 9646

Astragalis 9318

Astrebla 8861

Atriplex 8168, 8260, 8261,
8589, 8644, 8749, 8772, 8861,
8939, 8982, 8985, 9044, 9122,
9131, 9183, 9184, 9361, 9419,
9600, 9646

Atropa 9320

Avena 8156, 8168, 8182, 8245,
8428, 8472, 8745, 8755, 8833,
8835, 8864, 8951, 8991, 9099,
9120, 9121, 9131, 9299, 9375,
9389, 9461, 9462, 9523, 9571

Avicennia 9097, 9183

avocado see *Persea*

Azalea 8333

B

Bacteria
    Clostridium 9463
    Corynebacterium 9576
    Erwinia 8190, 8584, 8650
    Escherichia 9463
    Halobacterium 9284
    Pseudomonas 8190, 8862
    Rhizobium 8297, 8702,
    8904, 8945, 8946, 9155,
    9552, 9665
    Rhodospirillum 8865
    Streptococcus 9546

Balanops 8208

banana see *Musa*

Banksia 8579, 9529

barberry see *Berberis*

Barbeya 8208

barley see *Hordeum*

basswood see *Tilia*

Bauhinia 9302

bean see *Phaseolus*

beech see *Fagus*

Berberis 9422

bermudagrass see *Cynodon*

Beta 8145, 8147, 8172, 8173,
8182, 8190, 8199, 8244, 8305,
8307, 8313, 8325, 8427, 8488,
8503, 8513, 8516, 8545, 8614,
8673, 8685, 8719, 8742, 8772,
8805, 8861, 8957, 8985, 9034,
9058, 9146, 9200, 9218, 9223,
9252, 9266, 9297, 9382, 9419,
9429, 9430, 9431, 9518, 9551,
9646

Betula 8184, 8249, 8250, 8413,
8642, 8677, 8722, 8857, 9034,
9087, 9123, 9220, 9516

birch see *Betula*

blueberry see *Vaccinium*

Boerhaavia 9243

Borya 8772

Bouteloua 8160, 8293, 8294,
8804, 9321

Brachiaria 9021

Brassica 8180, 8182, 8287,
8307, 8346, 8393, 8394, 8454,
8516, 8535, 8557, 8614, 8655,
8804, 9095, 9098, 9208, 9252,
9282, 9297, 9620

broadleaf see *Hyptis*

brome grass see *Bromus*

Bromus 8182, 8465, 8466, 8607,
8644, 9193, 9252

Bryophyllum 8790, 9673

Bryophyta
    Anomodon 9210
    Bryum 8245
    Ceratodon 9210
    Conocephalum 9210
    Cratoneuron 8245, 8431
    Dicranum 9210
    Fissidens 8619
    Grimmia 9210
    Hylocomium 9412
    Leucobryum 9210
    Lunularia 9210
    Marchantia 9210
    Mnium 9210
    Neckera 8245
    Pellia 9210

Commelina 8158, 8441, 8442,
8533, 8654, 8745, 8940, 8941,
8942, 8943, 8947, 8948, 8950,
8951, 8952, 8953, 9139, 9230,
9231, 9232, 9294, 9295, 9312,
9359, 9429, 9543, 9618, 9626,
9640, 9641

Convallaria 8475

Corchorus 9104, 9138

Cordia 8872

Coriandrum 9371

cornel see *Cornus*

Cornus 8289, 8492, 8677, 8709,
8857, 9273, 9352

Coronilla 8857, 9318

Corydalis 8475

Corylus 8929, 9360

Cotoneaster 8131

cotton see *Gossypium*

cottonwood see *Populus*

cowpea see *Vigna*

cranberry see *Vaccinium*

Crataegus 8709

Crithmum 9103

Crocus 9144

Crotalaria 9243, 9314, 9318

Cryptocarya 9195

Cryptomeria 8590, 8742, 8857,
9081

cucumber see *Cucumis*

Cucumis 8182, 8240, 8303,
8321, 8325, 8344, 8545, 8631,
8755, 8971, 9098, 9252, 9277,
9477, 9540, 9551, 9571

Cucurbita 8182, 8394, 8635,
8711, 9037, 9429, 9535, 9571

currant see *Ribes*

Cyclamen 8877, 9626

Cyclanthera 9308

Cynara 8418

Cynodon 8182, 8347, 9021,
9252

Cyperus 8857, 8991, 9020,
9038, 9135

Cytisus 8929, 9318

D

Dactylis 8655, 8772, 8850,
9252

Datura 9463

Daucus 8303, 8490, 8588,
9098, 9103, 9252, 9282, 9371

Delphinium 8187

Deschampsia 8850, 8929, 9094

Descurainia 8688

Desmodium 8146, 8669, 8929,
9243

Dieffenbachia 9196

Digitaria 9021

Dionaea 9650

Diospyros 8455, 9244, 9578

Dipterocarpus 9159

Distichlis 8749, 8810, 8811,
8939

dogwood see *Cornus*

Dolichos 8557

Dorycnium 9318

Douglas fir see *Pseudotsuga*

Dryobalanops 9159

Dupontia 8929, 9099

SUBJECT INDEX

This index contains a selection of primary items chosen according to their interest
for water relation researchers and to their relative importance and occurrence.

A

Abaxial and adaxial leaf epidermes 8167, 8177, 8185, 8197, 8221, 8233, 8240, 8285,
     8309, 8461, 8467, 8475, 8557, 8579, 8645, 8686, 8699, 8744, 8775, 8791, 8818,
     8855, 8985, 8997, 8999, 9016, 9074, 9106, 9109, 9139, 9141, 9237, 9243, 9304,
     9318, 9372, 9373, 9374, 9382, 9422, 9443, 9543, 9585, 9597, 9600

Abscisic acid see Antitranspirants; Growth substances, hormones, inhibitors etc....

Absorption of water see Water absorption ...

Age of leaf see Age of plant ...; Leaf insertion level ...

Age of plant and conductance for water vapour and carbon dioxide transfer 8238,
     8250, 8460, 8537, 8627, 8729, 8768, 8843, 9329, 9408, 9443, 9475, 9494, 9499,
     9604

Age of plant and stomata 9329

Age of plant and transpiration 8252, 8390, 8407, 9326, 9375

Age of plant and water absorption by plant 8437

Age of plant and water status in plant 8134, 8184, 8303, 8627, 8648, 8662, 8700,
     8729, 8807, 9067, 9194, 9300, 9326, 9376, 9408, 9416, 9499

Age of plant and water transport in cells 9345

Age of plant and water transport in plant 9116, 9326

Age of plant and wilting 8313, 8367, 8627, 8653, 8893, 9200, 9431

Agrotechnics see Farming practices ...

Air conditioning see Transpiration rate, methods, gasometric systems, conditioning
     of air

Air-flow rate see Wind ...

Algae see Deuterium oxide, tritium oxide ...; Osmotically active substances ...;
     Salinity and productivity of algae

Altitude, pressure and transpiration 8216, 9011, 9061

Altitude, pressure and water status in plant 8216

Amino acids see Protein, amino acids, nucleic acids ...

Anatomical structure and conductance for water vapour and carbon dioxide transfer
     9175

Anatomical structure and transpiration 9175

Anatomical structure and water absorption by plant 8312

Anatomical structure and wilting   8285

Anatomical structure of epidermis 8397, 8747, 9349

Antibiotics and water status in plant   9535

Antitranspirants (see also Growth substances, hormones, inhibitors etc. ...) 8319, 8410, 8413, 8414, 8539, 8744, 8756, 8826, 8950, 8953, 9009, 9042, 9237, 9359, 9429, 9516, 9542, 9572, 9630, 9638

Assimilation see Carbon dioxide influx ...

Availability of soil water 8147, 8153, 8154, 8192, 8195, 8218, 8237, 8239, 8241, 8297, 8300, 8339, 8351, 8367, 8388, 8390, 8406, 8412, 8417, 8457, 8462, 8472, 8487, 8494, 8499, 8532, 8548, 8549, 8565, 8566, 8569, 8570, 8585, 8586, 8616, 8617, 8618, 8642, 8653, 8658, 8659, 8702, 8716, 8727, 8777, 8779, 8828, 8851, 8856, 8858, 8883, 8893, 8901, 8903, 8945, 8972, 8975, 9000, 9001, 9012, 9041, 9054, 9062, 9094, 9123, 9124, 9130, 9133, 9139, 9146, 9164, 9166, 9167, 9201, 9204, 9208, 9211, 9212, 9217, 9220, 9233, 9278, 9282, 9290, 9297, 9300, 9313, 9315, 9316, 9321, 9325, 9389, 9392, 9410, 9417, 9420, 9433, 9444, 9479, 9480, 9487, 9502, 9508, 9555, 9557, 9568, 9578, 9582, 9583, 9584, 9607, 9608

B

Beta gauge see Water saturation deficit, methods

Biological clock see Diurnal changes ...

Books on plant water relation see General aspects

Bound water 8225, 8235, 8689, 8805, 8855, 8976, 9011, 9067, 9139, 9248, 9250, 9309, 9363, 9379, 9471, 9659, 9664

Bound water, methods 8225, 8398

Boundary layer of leaf see Conductance for water vapour and carbon dioxide transfer, boundary layer of leaf

C

C_3, C_4, CAM pathways see Comparison of plants with different types of carbon metabolism

Canopy architecture see Drought ...; Humidity of air ...; Soil moisture ...; Water status in plant and canopy architecture

Canopy model see Model of canopy

Canopy water vapour profiles see Humidity of air, gradients in canopy

Carbohydrates see Saccharides ...

Carbon dioxide and conductance for water vapour and carbon dioxide transfer 8361, 8677, 8744, 8745, 8799, 8913, 8929, 8951, 8952, 8960, 9016, 9134, 9225, 9226, 9372, 9373, 9404

Carbon dioxide and stomata 8413, 8533, 8654, 8696, 8744, 8835, 8929, 8949, 8951, 8952, 8953, 9106, 9134, 9231, 9232, 9543, 9612, 9613, 9641

Conductance for water vapour and carbon dioxide transfer, canopy 8332, 8528, 8637,
     8740, 8743, 8744, 8840, 8912, 9064, 9152, 9164, 9290, 9443, 9569, 9607

Conductance for water vapour and carbon dioxide transfer, cuticle 8743, 8857, 9261,
     9370

Conductance for water vapour and carbon dioxide transfer, diurnal changes 8133,
     8140, 8141, 8197, 8204, 8216, 8221, 8238, 8239, 8262, 8359, 8374, 8411, 8445,
     8457, 8460, 8494, 8525, 8538, 8540, 8644, 8645, 8662, 8667, 8729, 8744, 8773,
     8776, 8801, 8814, 8823, 8883, 8838, 8839, 8850, 8851, 8855, 8860, 8907, 8912,
     8913, 8950, 8951, 8999, 9041, 9064, 9066, 9082, 9087, 9094, 9134, 9139, 9141,
     9152, 9166, 9195, 9253, 9280, 9297, 9329, 9360, 9372, 9422, 9443, 9496, 9497,
     9498, 9514, 9515, 9534, 9570, 9599, 9606, 9622, 9623, 9628, 9660, 9661

Conductance for water vapour and carbon dioxide transfer, genetics 8309, 8460, 8461,
     8831, 9054, 9141

Conductance for water vapour and carbon dioxide transfer, heterogeneity of single
     leaf blade 8233

Conductance for water vapour and carbon dioxide transfer, intercellular spaces 8576,
     8857, 8861, 9134

Conductance for water vapour and carbon dioxide transfer, oscillations 9661

Conductance for water vapour and carbon dioxide transfer, seasonal changes 8197,
     8250, 8289, 8385, 8460, 8512, 8615, 8667, 8676, 8727, 8921, 9041, 9152, 9195,
     9209, 9306, 9321, 9360, 9494, 9498, 9514, 9606, 9607

Conductance for water vapour and carbon dioxide transfer, stomata 8130, 8131, 8133,
     8134, 8137, 8140, 8141, 8159, 8167, 8189, 8190, 8192, 8197, 8198, 8204, 8216,
     8221, 8224, 8233, 8239, 8240, 8244, 8249, 8250, 8251, 8252, 8260, 8262, 8263,
     8265, 8284, 8289, 8290, 8296, 8306, 8309, 8315, 8325, 8326, 8332, 8333, 8334,
     8359, 8360, 8361, 8374, 8383, 8385, 8387, 8406, 8411, 8413, 8416, 8417, 8433,
     8445, 8449, 8457, 8460, 8461, 8474, 8494, 8495, 8531, 8536, 8537, 8538, 8540,
     8546, 8549, 8590, 8591, 8603, 8606, 8615, 8627, 8641, 8644, 8645, 8662, 8664,
     8665, 8667, 8670, 8677, 8710, 8727, 8729, 8740, 8742, 8743, 8744, 8745, 8750,
     8751, 8768, 8769, 8770, 8776, 8799, 8800, 8801, 8809, 8810, 8811, 8813, 8814,
     8818, 8823, 8826, 8831, 8833, 8838, 8839, 8857, 8860, 8861, 8891, 8894, 8895,
     8912, 8913, 8914, 8921, 8923, 8925, 8926, 8927, 8929, 8944, 8949, 8951, 8952,
     8953, 8960, 8985, 8997, 8999, 9007, 9016, 9032, 9033, 9034, 9041, 9042, 9050,
     9054, 9066, 9071, 9074, 9086, 9096, 9109, 9110, 9129, 9134, 9141, 9157, 9158,
     9160, 9164, 9166, 9178, 9209, 9225, 9237, 9250, 9253, 9261, 9267, 9273, 9278,
     9280, 9288, 9290, 9297, 9306, 9321, 9327, 9337, 9360, 9372, 9376, 9395, 9397,
     9401, 9402, 9404, 9407, 9408, 9422, 9428, 9435, 9443, 9444, 9445, 9460, 9475,
     9494, 9496, 9517, 9518, 9522, 9523, 9534, 9550, 9569, 9570, 9571, 9577, 9582,
     9599, 9604, 9606, 9622, 9623, 9627, 9628, 9639, 9656, 9659, 9660 9661

Conductance for water vapour transfer, epidermis 8131, 8159, 8160, 8198, 8221, 8231,
     8232, 8238, 8240, 8250, 8252, 8261, 8262, 8290, 8296, 8307, 8332, 8334, 8360,
     8365, 8374, 8387, 8406, 8423, 8445, 8457, 8470, 8475, 8509, 8512, 8532, 8576,
     8628, 8629, 8662, 8667, 8673, 8676, 8719, 8727, 8736, 8740, 8750, 8751, 8763,
     8771, 8773, 8801, 8813, 8818, 8821, 8839, 8840, 8843, 8849, 8850, 8851, 8855,
     8857, 8861, 8891, 8902, 8907, 8925, 8951, 8997, 8999, 9001, 9007, 9034, 9074,
     9082, 9085, 9087, 9094, 9123, 9129, 9134, 9139, 9175, 9178, 9195, 9210, 9224,
     9226, 9239, 9249, 9250, 9261, 9274, 9278, 9290, 9321, 9329, 9334, 9337, 9360,
     9369, 9372, 9373, 9395, 9422, 9514, 9515, 9523, 9528, 9529, 9530, 9579, 9582,
     9623, 9625, 9628, 9639, 9644, 9660

Consumption of water see Water consumption

Crasulacean Acid Metabolism see Comparison of plants with different types of carbon
     metabolism

Cryoscopy see Osmotic potential, methods

Cultivars and conductance for water vapour and carbon dioxide transfer 8167, 8334,
      8378, 8457, 8512, 8645, 8667, 8736, 8823, 8944, 9071, 9074, 9239, 9329, 9397,
      9401, 9402, 9428

Cultivars and stomata 8177, 8582, 8645, 9069, 9239

Cultivars and transpiration 8407, 8605, 8667, 9186, 9653, 9654

Cultivars and water absorption by plant 8312, 9651, 9654

Cultivars and water status in plant 8457, 8482, 8645, 8653, 8667, 8944, 8988, 9067,
      9134, 9336, 9381, 9401, 9428

Cultivars and water transport in plant 8375

Cultivars and wilting 8482, 8653, 8927, 9336

Cuticle see Conductance for water vapour and carbon dioxide transfer, cuticle

Cuticular transpiration see Transpiration, cuticular

D

Decapitation see Defoliation, decapitation, ear, root removal ...

Defoliation, decapitation, ear, root removal and conductance for water vapour and
      carbon dioxide transfer 8537, 8670

Defoliation, decapitation, ear, root removal and transpiration 8507, 8670, 8904

Defoliation, decapitation, ear, root removal and water absorption by plant 9211

Defoliation, decapitation, ear, root removal and water status in plant 9211

Desiccation see Water saturation deficit; Wilting ...

Deuterium oxide, tritium oxide and biliproteins 9674

Deuterium oxide, tritium oxide and carbon dioxide influx 8492

Deuterium oxide, tritium oxide and chlorophyll 8848

Deuterium oxide, tritium oxide and electron transport chain 8852

Deuterium oxide, tritium oxide and productivity of algae 8573

Deuterium oxide, tritium oxide and transpiration 9327

Deuterium oxide, tritium oxide and water absorption by plant 8881, 9079

Deuterium oxide, tritium oxide and water status in plant 8492

Deuterium oxide, tritium oxide and water transport in cells 9284

Deuterium oxide, tritium oxide and water transport in plant 8955, 9346

Deuterium oxide, tritium oxide, methods 8695

Deuterium oxide, tritium oxide, occurrence 9465, 9466

Dew   see   Precipitation, dew ...

Dew point hygrometer   see   Water potential, methods, dew point hygrometer

Diffusion coefficient 9034

Diffusion (diffusive) conductance see Conductance ...

Diffusion porometers see Stomatal aperture, methods, diffusion porometers

Diffusion (diffusive) resistance see Conductance ...

Disease see Pathogens ...

Diurnal changes see Conductance for water vapour and carbon dioxide transfer, ...;
        Stomatal aperture, ...; Transpiration rate, ...; Water absorption by plant,
        ...; Water status in plant, ...; Water transport in plant, ...; Wilting,
        diurnal changes

D_2O, T_2O see Deuterium oxide, tritim oxide ...

Drainage see Farming practices ...

Drought and canopy architecture 8317

Drought and carbon dioxide influx 8371, 8467, 8506, 8627, 8861, 8901, 8922, 8981,
        9044, 9134, 9506, 9583, 9605

Drought and carbon fixation pathways 8467, 8973, 9406

Drought and chlorophyll 8857, 8973, 9057, 9134, 9188, 9385

Drought and chloroplasts 9134, 9188, 9323

Drought and conductance for water vapour and carbon dioxide transfer 8406, 8413,
        8627, 8645, 8667, 8744, 8809, 8840, 9134, 9141, 9404, 9506

Drought and electron transport chain 8973, 9385

Drought and growth, productivity 8134, 8202, 8243, 8278, 8294, 8317, 8347, 8362,
        8403, 8404, 8409, 8433, 8435, 8494, 8506, 8511, 8548, 8552, 8554, 8565, 8604,
        8608, 8613, 8624, 8626, 8627, 8653, 8671, 8684, 8712, 8727, 8751, 8761, 8768,
        8774, 8809, 8901, 8927, 8939, 8962, 8994, 9019, 9029, 9071, 9098, 9118, 9123,
        9132, 9133, 9166, 9234, 9271, 9301, 9325, 9392, 9404, 9453, 9480, 9483, 9493,
        9502, 9506, 9557, 9581, 9583, 9587, 9608

Drought and leaf anatomy 8129, 8280, 8471, 8644, 8809, 8844, 8901, 9002, 9081, 9506

Drought and photorespiration 8861

Drought and respiration 9173

Drought and stomata 8413, 8744, 9141

Drought and transpiration 8171, 8277, 8506, 8605, 8981, 9044, 9479, 9483, 9506,
        9621

Drought and water absorption by plant 8218, 9479

Drought and water status in plant 8153, 8239, 8241, 8406, 8840, 8962, 9141, 9392,
        9450

Drought and water transport in cells 9134, 9188

Hydraulic conductivity see Water transport in plant ...

Hydroactive closure of stomata see Water status in plant and stomata

Hydrogen isotopes see Deuterium oxide, tritium oxide ...

Hygrometer see Humidity of air, methods

Hypostomatous leaves 9306, 9422

I

Ideotype see Model ...

Infra-red gas analysers see Transpiration rate, methods, infra-red gas analysers

Inhibitors see Growth substances, hormones, inhibitors etc. ...

Insertion level see Leaf insertion level ...

Integrated transpiration see Transpiration, integrated

Intercellular spaces see Conductance for water vapour and carbon dioxide transfer,
        intercellular spaces; Transpiration rate, gradients of air humidity in leaf
        intercellular spaces

Irradiance and conductance for water vapour and carbon dioxide transfer 8141, 8197,
        8238, 8307, 8334, 8359, 8361, 8417, 8460, 8509, 8576, 8590, 8662, 8673, 8676,
        8677, 8729, 8742, 8744, 8745, 8799, 8801, 8810, 8811, 8838, 8912, 8913, 8914,
        8926, 8929, 8950, 8951, 8952, 8999, 9064, 9066, 9129, 9134, 9141, 9164, 9193,
        9372, 9373, 9376, 9443, 9496, 9497, 9499, 9514, 9550, 9571, 9628, 9644, 9656,
        9659, 9660, 9661

Irradiance and stomata 8442, 8654, 8696, 8744, 8758, 8915, 8929, 8950, 8951, 8952,
        9101, 9110, 9128, 9134, 9231, 9232, 9294, 9295, 9304, 9372, 9373, 9543, 9571,
        9667

Irradiance and transpiration 8171, 8252, 8270, 8359, 8485, 8590, 8729, 8764, 8765,
        8801, 8838, 8861, 8896, 8898, 8909, 8999, 9036, 9044, 9090, 9101, 9122, 9129,
        9134, 9193, 9216, 9452, 9514, 9638, 9644, 9660

Irradiance and water absorption by plant 8297, 9079

Irradiance and water status in plant 8136, 8159, 8162, 8549, 8709, 8715, 8778, 8810,
        8811, 8816, 8915, 9002, 9122, 9126, 9141, 9447, 9497, 9499, 9622, 9644, 9660

Irradiance and water transport in plant 8159

Irradiance and wilting 9063, 9122, 9141

Irrigation and carbon dioxide influx 8186, 8227, 8520, 8662, 8901, 8923, 8961, 9267,
        9338, 9360

Irrigation and chlorophyll 9409

Irrigation and conductance for water vapour and carbon dioxide transfer 8140, 8378,
        8416, 8457, 8662, 8727, 8744, 9267, 9401, 9408, 9428, 9443

Irrigation and growth, productivity 8138, 8139, 8140, 8145, 8172, 8173, 8176, 8179,
        8186, 8193, 8206, 8217, 8273, 8275, 8287, 8303, 8305, 8335, 8336, 8341, 8344,
        8346, 8351, 8368, 8374, 8378, 8384, 8388, 8393, 8394, 8409, 8416, 8417, 8438,
        8450, 8454, 8465, 8466, 8484, 8486, 8500, 8513, 8520, 8541, 8542, 8551, 8554,

Mutagens, other organic substances and stomata 8190, 8387, 8744, 9009, 9359, 9648

Mutagens, other organic substances and transpiration 8168, 8319, 9429, 9648

Mutagens, other organic substances and water absorption by plant 9509

Mutagens, other organic substances and water status in plant 8326, 8360, 8387, 9009, 9134, 9647

Mutagens, other organic substances and wilting 8350, 8387, 8844, 9134, 9158, 9178, 9647

Mutants and transpiration 9653

N

Nitrogen see Mineral elements ...

Nucleic acids see Proteins, amino acids, nucleic acids ...

O

O$_2$ see Oxygen ...

O$_3$ see Pollutants, ozone ...

Ontogeny see Age of plant ...

Oscillations see Conductance for water vapour and carbon dioxide transfer, ...; Transpiration rate, ...; Wilting, oscillations

Osmotic potential in plant tissue (see also Water status in plant ...) 8130, 8133, 8151, 8152, 8153, 8188, 8189, 8190, 8219, 8241, 8259, 8282, 8328, 8333, 8349, 8351, 8379, 8398, 8401, 8417, 8424, 8434, 8435, 8448, 8449, 8494, 8497, 8522, 8575, 8588, 8636, 8674, 8675, 8677, 8688, 8699, 8708, 8715, 8716, 8744, 8749, 8750, 8751, 8758, 8772, 8783, 8799, 8800, 8803, 8807, 8816, 8821, 8826, 8827, 8830, 8857, 8868, 8875, 8877, 8906, 8907, 8928, 8931, 8934, 8939, 8940, 8956, 8970, 8989, 8992, 8998, 9002, 9009, 9031, 9052, 9059, 9063, 9123, 9134, 9151, 9168, 9173, 9183, 9195, 9213, 9250, 9252, 9261, 9273, 9296, 9330, 9331, 9343, 9344, 9357, 9361, 9376, 9404, 9416, 9418, 9434, 9436, 9461, 9462, 9497, 9499, 9529, 9531, 9538, 9556, 9558, 9559, 9560, 9561, 9562, 9564, 9580, 9602, 9604, 9615, 9622, 9623, 9644, 9646, 9647, 9661, 9670

Osmotic potential in substrate 8228, 8274, 8329, 8343, 8372, 8402, 8418, 8453, 8548 8548, 8675, 8721, 8750, 8772, 8779, 8795, 8839, 8878, 8944, 8977, 9004, 9093, 9096, 9119, 9343, 9521, 9646

Osmotic potential, methods 9344, 9556

Osmotically active substances and carbon dioxide influx 8228, 8242, 8750, 8780, 878 8781, 8782, 8815, 8829, 8932, 9593, 9652

Osmotically active substances and carbon fixation pathways 8698, 8780, 8783, 8986, 9335

Osmotically active substances and carotenoids 9540

Osmotically active substances and chlorophyll 8210, 8646, 9335

Osmotically active substances and chloroplasts 8419, 8754, 8780, 8781, 8782, 8783, 8932, 9309, 9310, 9311

Potometry 8840, 8841, 8980, 9296

Precipitation, dew and carbon dioxide influx 9011

Precipitation, dew and carbon fixation pathways 9624

Precipitation, dew and growth, productivity 8278, 8346, 8373, 8404, 8427, 8440,
     8455, 8477, 8511, 8524, 8552, 8571, 8591, 8604, 8656, 8680, 8734, 8738, 8769,
     8938, 8978, 8979, 9011, 9124, 9319, 9420, 9433, 9493

Precipitation, dew and leaf anatomy 8471, 8656

Precipitation, dew and transpiration 8236, 8330, 9152, 9555

Precipitation, dew and water absorption by plant 6569, 8858, 9480

Precipitation, dew and water status in plant 8629, 9422

Precipitation, dew and water transport in plant

Precipitation, dew, methods 8166, 8559, 8623, 9619

Pressure bomb see Water potential, methods, pressure bomb

Pressure potential in plant tissue (see also Water status in plant ...) 8132,
     8133, 8134, 8153, 8155, 8162, 8188, 8189, 8241, 8256, 8258, 8259, 8301, 8349,
     8376, 8377, 8379, 8386, 8398, 8399, 8401, 8413, 8417, 8448, 8449, 8489, 8490,
     8494, 8575, 8612, 8621, 8666, 8676, 8677, 8706, 8715, 8716, 8743, 8744, 8749,
     8750, 8751, 8800, 8826, 8827, 8857, 8905, 8931, 8932, 8939, 8940, 8956, 8970,
     8976, 8981, 8989, 8998, 9002, 9009, 9031, 9063, 9123, 9134, 9151, 9176, 9177,
     9188, 9195, 9261, 9273, 9330, 9376, 9404, 9434, 9436, 9444, 9456, 9457, 9458,
     9499, 9529, 9531, 9539, 9556, 9558, 9559, 9562, 9564, 9604, 9644, 9646, 9659,
     9661, 9675

Pressure potential, methods 8621, 8706, 9434, 9444

Productivity see Growth, productivity ...

Productivity of algae see Osmotically active substances ...; Salinity and pro-
     ductivity of algae

Productivity of transpiration 8175, 8192, 8203, 8204, 8205, 8238, 8278, 8279, 8311,
     8325, 8409, 8465, 8536, 8576, 8613, 8618, 8764, 8813, 8855, 8857, 8860, 8861,
     8909, 8951, 9011, 9090, 9129, 9134, 9139, 9186, 9217, 9429, 9500, 9510, 9514,
     9516, 9567, 9568, 9606, 9642

Proteins, amino acids, nucleic acids and stomata 9237

Proteins, amino acids, nucleic acids and water status in plant 8182, 8245, 8674,
     9134, 9477

Proteins, amino acids, nucleic acids and water transport in plant 9573

Proteins, amino acids, nucleic acids and wilting 8182, 8245, 9134, 9463, 9617

Psychrometry see Water potential, methods, psychrometry

P/T ratio see Productivity of transpiration

R

Radiation see Irradiance ...

Rain see Precipitation, dew ...

Reactivity of stomata see Stomatal reaction rate; Stomatal reactivity during
      ontogeny

Rehydration 8131, 8156, 8182, 8239, 8245, 8271, 8294, 8296, 8345, 8367, 8413, 8431,
      8448, 8488, 8515, 8522, 8589, 8641, 8665, 8744, 8751, 8772, 8782, 8818, 8821,
      8843, 8861, 8874, 8899, 8907, 8911, 8932, 8951, 8954, 8959, 8994, 9015, 9063,
      9108, 9134, 9159, 9176, 9177, 9278, 9317, 9350, 9353, 9354, 9366, 9390, 9403,
      9404, 9506, 9520, 9563, 9602, 9647, 9659

Relative water content see Water saturation deficit

Resistance see Conductance ...

Respiration see Drought ...; Flooding ...; Humidity of air ...; Osmotically
      active substances ...;   Salinity ...; Soil moisture ...; Water content
      in plant and respiration

Root pressure, exudation 8292, 8436, 8437, 8504, 8577, 8778, 9052, 9182, 9184,
      9436, 9636, 9670

Root removal see Defoliatin, decapitation, ear, root removal ...

Root, underground part and growth, productivity 8292, 8354, 8367, 8381, 8414, 8548,
      8565, 8796, 8972, 8983, 9140, 9271, 9297, 9391, 9525, 9618, 9623

Root, underground part and water absorption by plant 8472, 8527, 9153

Root, underground part and wilting 8772, 9134

S

Saccharides and stomata 8405

Saccharides and transpiration 8480

Saccharides and water status in plant 8152, 8182, 8482, 8636, 8708, 8772, 9134

Saccharides and water transport in cells 8808

Saccharides and wilting 8429, 8482, 8522, 8622, 8636, 9617

Saline water see Irrigation water quality

Salinity and carbon dioxide influx 8130, 8228, 8264, 8342, 8415, 8587, 8657, 8660,
      8673, 8674, 8750, 8810, 8811, 8815, 8829, 8886, 9604

Salinity and carbon fixation pathways 8164, 8673, 8674, 9107, 9596, 9597, 9598,
      9604

Salinity and carotenoids 8657, 9242, 9602

Salinity and chlorophyll 8144, 8145, 8164, 8342, 8351, 8380, 8646, 8657, 9242,
      9322, 9596, 9598

Salinity and conductance for water vapour and carbon dioxide transfer 8130, 8449,
      8673, 8810, 8811, 9604

Soil moisture and growth, productivity (continued) 9017, 9028, 9029, 9056, 9073,
     9074, 9131, 9133, 9134, 9138, 9140, 9145, 9146, 9147, 9155, 9170, 9171, 9183,
     9199, 9200, 9201, 9204, 9214, 9215, 9217, 9218, 9220, 9222, 9247, 9262, 9269,
     9271, 9275, 9283, 9292, 9297, 9298, 9301, 9303, 9313, 9315, 9320, 9333, 9377,
     9384, 9392, 9398, 9399, 9400, 9410, 9413, 9414, 9415, 9420, 9430, 9432, 9433,
     9453, 9467, 9482, 9483, 9488, 9506, 9507, 9519, 9524, 9529, 9545, 9558, 9567,
     9578, 9582, 9583, 9586, 9591, 9608, 9618, 9634, 9669

Soil moisture and leaf anatomy 8280, 8363, 8395, 8411, 8933, 8959, 9001, 9074, 9114
     9114, 9292

Soil moisture and photorespiration 8719

Soil moisture and respiration 8719, 9090

Soil moisture and stomata 8413, 8744, 8768, 8857, 9074, 9110, 9382, 9422

Soil moisture and transpiration 8160, 8198, 8207, 8237, 8252, 8263, 8267, 8352,
     8383, 8390, 8625, 8629, 8672, 8751, 8799, 8857, 8959, 8981, 8999, 9001, 9044,
     9066, 9090, 9129, 9139, 9152, 9278, 9283, 9326, 9364, 9397, 9483, 9506, 9507,
     9516, 9536, 9615, 9642, 9648, 9654

Soil moisture and water absorption by plant 8218, 8392, 8417, 9262, 9654

Soil moisture and water status in plant 8160, 8280, 8304, 8406, 8444, 8457, 8549,
     8629, 8658, 8672, 8701, 8709, 8714, 8715, 8857, 8874, 8888, 8962, 8999, 9001,
     9126, 9141, 9170, 9195, 9212, 9283, 9300, 9321, 9326, 9422, 9519, 9528, 9529,
     9555, 9586, 9601

Soil moisture and water transport in plant 8267, 9326

Soil moisture and wilting 8406, 8701, 8751, 8874, 8893, 9141, 9262, 9336

Soil moisture control, methods 9276, 9333, 9431

Soil moisture, methods 8234, 8581, 8785, 9062, 9290, 9444

Soil water potential see Water potential in substrate

Solar radiation see Irradiance ...

Stomata see Age of plant ...; Carbon dioxide ...; Conductance for water vapour
     and carbon dioxide transfer ...; Cultivars ...; Drought ...; Ecotypes,
     geographical types ...; Enzyme inhibitors ...; Flooding ...; Genetics ...;
     Growth substances, hormones, inhibitors etc. ...; Humidity of air ...;
     Irradiance ...; Leaf insertion level ...; Mineral elements ...; Mutagens,
     other organic substances ...; Osmotically active substances ...; Oxygen ...;
     Pathogens ...; Pollutants, ozone ...; Proteins, amino acids, nucleic acids
     ...; Saccharides ...; Salinity ...; Soil moisture ...; Taxons ...; Tem-
     perature ...; Water status in plant ...; Wind and stomata

Stomata and epidermis, heterogeneity of single leaf blade 8948, 9016

Stomata and photosynthetic rate 8224, 8238, 8239m 8261, 8306, 8470, 8515, 8525,
     8740, 8742, 8744, 8745, 8857, 8861, 8891, 8894, 8913, 8953, 8991, 9041, 9134,
     9360, 9551, 9582

Stomata and transpiration rate 8171, 8239, 8590, 8591, 8729, 8740, 8744, 8745, 8800,
     8857, 8953, 8991, 8999, 9066, 9074, 9134, 9141, 9243, 9429, 9443, 9510, 9523,
     9551

Stomata development 8180, 8187, 8620, 8744, 8766, 8819, 8991, 9020, 9135, 9136,
     9144, 9245, 9246, 9371, 9667

Structure of epidermis 8161, 8185, 8215, 8285, 8323, 8397, 8405, 8579, 8819, 8933, 9016, 9144, 9159, 9185, 9304

Sulphur oxides and other sulphur compounds see Pollutants, ozone ...

Surface impressions see Stomatal aperture, methods, microrelief methods

T

Taxons and conductance for water vapour and carbon dioxide transfer 8140, 8325, 8850, 8999, 9087, 9373, 9395

Taxons and stomata 8208, 8654, 8850, 9668

Taxons and transpiration 8999, 9001, 9395, 9516

Taxons and water status in plant 8140, 8422, 8850, 8999, 9123, 9322, 9617

Taxons and wilting 8681

Temperature and conductance for water vapour and carbon dioxide transfer 8130, 8252, 8260, 8307, 8417, 8460, 8509, 8576, 8677, 8744, 8745, 8787, 8799, 8801, 8810, 8811, 8860, 8894, 8895, 8913, 8929, 8960, 9007, 9032, 9064, 9085, 9166, 9460, 9494, 9514, 9625, 9628

Temperature and stomata 8699, 8744, 8758, 8929, 9285, 9588

Temperature and transpiration 8171, 8252, 8270, 8315, 8393, 8485, 8576, 8591, 8625, 8764, 8801, 8836, 8909, 8935, 8960, 9033, 9044, 9122, 9290, 9395, 9429, 9514, 9625

Temperature and water absorption by plant 8881, 8910, 9052, 9553, 9636

Temperature and water status in plant 8200, 8201, 8213, 8400, 8553, 8699, 8709, 8784, 8806, 8810, 8811, 8960, 8987, 8988, 9096, 9122, 9228, 9447, 9601

Temperature and water transport in cells 8832, 9332, 9356, 9455, 9472, 9526, 9539, 9574

Temperature and water transport in plant 8787, 9495, 9636

Temperature and wilting 8554, 9122, 9336

Transpiration see Age of plant ...; Altitude, pressure ...; Anatomical structure ...; Carbon dioxide ...; Cultivars ...; Defoliation, decapitation, ear, root removal ...; Deuterium oxide, tritium oxide ...; Drought ...; Ecotypes ...; Farming practices ...; Flooding ...; Genetics ...; Growth substances, hormones, inhibitors etc. ...; Humidity of air ...; Irradiance ...; Irrigation ...; Leaf insertion level ...; Lipids, fatty acids ...; Mineral elements ...; Mutagens, other organic substances ...; Mutants ...; Osmotically active substances ...; Oxygen ...; Pathogens ...; Pesticides, herbicides ...; Photoperiod ...; Pollutants, ozone ...; Precipitation, dew ...; Saccharides ...; Salinity ...; Soil moisture ...; Taxons ...; Temperature ...; Water status in plant ...; Wind and transpiration

Transpiration chambers 8231, 8232, 8270, 8470, 8473, 8540, 8598, 8793, 8798, 8801, 8913, 8919, 8921, 8965, 9112, 9338, 9388

Transpiration coefficient 8130, 8139, 8154, 8222, 8278, 8330, 8334, 8390, 8394, 8487, 8511, 8617, 8681, 8723, 8747, 8828, 8857, 8860, 9040, 9054, 9074, 9090, 9200, 9217, 9233, 9262, 9313, 9360, 9364, 9375, 9386, 9396, 9416, 9446, 9488, 9510, 9536, 9567, 9582, 9583, 9606, 9607, 9660

Transpiration curves 8772, 9134

Transpiration cuticular 8516, 9241, 9243, 9244, 9254, 9365

Transpiration integrated 8693, 8729, 8840, 8944, 9074, 9262, 9337, 9516, 9583, 9654

Transpiration rate and antitranspirants see Antitranspirants ...

Transpiration rate and leaf temperature 8591, 9033, 9046, 9268

Transpiration rate and stomata see Stomata and transpiration rate

Transpiration rate, comparison of plants with different types of carbon metabolism
    8168

Transpiration rate, diurnal changes 8198, 8207, 8216, 8218, 8219, 8238, 8252, 8262,
    8330, 8359, 8435, 8538, 8682, 8688, 8709, 8729, 8731, 8740, 8743, 8764, 8790,
    8801, 8838, 8841, 8857, 8860, 8899, 8913, 8919, 8921, 8925, 8965, 8999, 9009,
    9035, 9036, 9043, 9044, 9090, 9152, 9253, 9439, 9452, 9514, 9515, 9628, 9642,
    9648, 9654, 9660

Transpiration rate, gradients of air humidity in leaf intercellular spaces 8857,
    9047, 9192

Transpiration rate in artificial conditions 8130, 8131, 8160, 8168, 8252, 8267,
    8325, 8333, 8372, 8381, 8383, 8407, 8429, 8475, 8479, 8507, 8517, 8526, 8539,
    8576, 8646, 8667, 8670, 8682, 8744, 8751, 8778, 8792, 8801, 8838, 8846, 8898,
    8904, 8921, 8935, 8955, 8959, 8981, 8985, 8992, 9009, 9040, 9072, 9101, 9113,
    9129, 9162, 9186, 9193, 9210, 9253, 9261, 9326, 9342, 9360, 9423, 9443, 9506,
    9516, 9551, 9580, 9615, 9623, 9625, 9628, 9638, 9642, 9648, 9660

Transpiration rate in natural conditions 8171, 8175, 8209, 8216, 8218, 8219, 8238,
    8262, 8263, 8270, 8300, 8330, 8352, 8359, 8383, 8390, 8506, 8510, 8518, 8538,
    8605, 8629, 8630, 8641, 8642, 8645, 8651, 8662, 8672, 8688, 8707, 8729, 8743,
    8765, 8840, 8841, 8855, 8897, 8899, 8913, 8919, 8925, 8981, 8994, 9009, 9011,
    9013, 9035, 9044, 9066, 9072, 9082, 9090, 9166, 9195, 9216, 9243, 9267, 9283,
    9297, 9327, 9328, 9440, 9483, 9491, 9499, 9501, 9512, 9514, 9515, 9579, 9583,
    9627, 9634, 9644, 9654, 9671

Transpiration rate, methods, gasometric systems, conditioning of air 8270

Transpiration rate, methods, gasometric systems, generally 8270, 9569

Transpiration rate, methods, gasometric systems, open 8473, 8913, 8921, 8923

Transpiration rate, methods, gasometric systems, semiclosed and closed 8470, 8603

Transpiration rate, methods, gravimetric 8510, 9047

Transpiration rate, methods, infra-red gas analysers 8913, 8921

Transpiration rate, methods, other hygrometers 8965

Transpiration rate, oscillations 8860, 9129, 9565

Transpiration rate, seasonal changes 8218, 8263, 8311, 8330, 8390, 8456, 8493, 8494,
    8562, 8629, 8659, 8765, 8825, 8851, 8857, 8925, 9011, 9012, 9035, 9041, 9061,
    9072, 9082, 9152, 9164, 9191, 9195, 9328, 9446, 9452, 9453, 9478, 9508, 9514,
    9555, 9606, 9607, 9654

Transpiration rate, theoretical background 8203, 8204, 8262, 8315, 8316, 8494, 8517,
    8591, 8630, 8638, 8731, 8732, 8740, 8857, 8891, 8925, 9033, 9036, 9047, 9048,
    9061, 9064, 9075, 9089, 9167, 9192, 9288, 9438, 9439, 9446, 9452, 9565, 9627,
    9644

Transpiration stomatal 9243, 9244, 9254

Transport of water see Water transport ...

Trichomes see Leaf surface, waxes and trichomes

Turgor pressure see Pressure potential ...

V

Vapour pressure deficit see Humidity of air ...

Virus diseases see Pathogens ...

W

Water absorption by parts of plant 9357, 9540

Water absorption by plant see Age of plant ...; Anatomical structure ...; Cul-
     tivars ...; Defoliation, decapitation, ear, root removal ...; Deuterium
     oxide, tritium oxide ...; Drought ...; Enzyme inhibitors ...; Enzymes ...;
     Farming practices ...; Flooding ...; Growth substances, hormones, inhibitors
     etc. ...; Humidity of air ...; Irradiance ...; Irrigation ...; Lipids,
     fatty acids ...; Mineral elements ...; Osmotically active substances ...;
     Oxygen ...; Pathogens ...; Precipitation, dew ...; Salinity ...; Soil
     moisture ...; Temperature and water absorption by plant

Water absorption by plant and ion uptake 8321, 8354, 8421, 8586, 8868, 9005, 9037,
     9134, 9183, 9266, 9305, 9436, 9531, 9541, 9600

Water absorption by plant, diurnal changes 8840, 8857, 9043, 9654

Water absorption by plant, seasonal changes 8825, 8857, 9211, 9557, 9568, 9654

Water absorption by seed 8183, 8274, 8312, 8392, 8402, 8469, 8498, 8501, 8550, 8610,
     8847, 8881, 8910, 8937, 9070, 9076, 9120, 9121, 9138, 9203, 9314, 9363, 9390,
     9391, 9553, 9554, 9643, 9651

Water absorption from atmosphere 8339, 8672, 9060

Water absorption from soil 8143, 8267, 8292, 8339, 8390, 8416, 8417, 8472, 8527,
     8569, 8630, 8651, 8653, 8677, 8727, 8732, 8743, 8857, 8925, 9037, 9123, 9140,
     9266, 9297, 9396, 9509, 9541, 9582, 9627, 9654

Water absorption from solution 8159, 8308, 8840, 8898, 9079, 9296

Water balance of cells and tissues 8974

Water balance of whole plant 8133, 8390, 8711, 8743, 8898, 8904, 9139, 9152, 9154,
     9191, 9541, 9558, 9583

Water consumption 8139, 8154, 8195, 8275, 8287, 8313, 8409, 8472, 8511, 8526, 8617,
     8681, 8723, 8727, 8774, 8826, 8994, 8999, 9054, 9200, 9208, 9233, 9297, 9300,
     9313, 9375, 9396, 9408, 9428, 9430, 9431, 9488, 9516, 9568, 9572, 9589, 9594,
     9671

Water consumption, methods 8613, 8703, 9051, 9507

Water saturation deficit 8241, 8247, 8271, 8289, 8304, 8360, 8379, 8398, 8406, 8475,
     8476, 8544, 8666, 8677, 8701, 8709, 8715, 8716, 8743, 8750, 8751, 8770, 8771,
     8772, 8787, 8821, 8831, 8843, 8857, 8874, 8877, 8927, 8951, 8954, 8979, 8989,
     8992, 9005, 9057, 9074, 9095, 9119, 9123, 9134, 9151, 9159, 9207, 9249, 9250,
     9261, 9273, 9283, 9292, 9293, 9316, 9317, 9329, 9342, 9392, 9393, 9403, 9404,
     9499, 9520, 9556, 9564, 9617, 9627, 9659

Water saturation deficit, methods 8715

Water status in plant see also Age of plant ...; Altitude, pressure ...; Anti-
     biotics ...; Carbon dioxide ...; Cultivars ...; Defoliation, decapitation,
     ear, root removal ...; Deuterium oxide, tritium oxide ...; Drought ...;
     Ecotypes, geographical types ...; Enzymes ...; Farming practices ...;
     Flooding ...; Genetics ...; Growth substances, hormones, inhibitors etc.
     ...; Humidity of air ...; Irradiance ...; Irrigation ...; Leaf insertion
     level ...; Mineral elements ...; Mutagens, other organic substances ...;
     Osmotically active substances ...; Oxygen ...; Pathogens ...; Pesticides,
     herbicides ...; pH ...; Photoperiod ...; Pollutants, ozone ...; Precipi-
     tation, dew ...; Proteins, amino acids, nucleic acids ...; Saccharides ...;
     Salinity ...; Soil moisture ...; Taxons ...; Temperature ...; Wind and
     water status in plant

Water status in plant and canopy architecture 8474, 8732, 9122

Water status in plant and carbon dioxide influx 8132, 8133, 8189, 8205, 8216, 8224,
     8228, 8239, 8242, 8306, 8352, 8445, 8470, 8596, 8597, 8598, 8619, 8677, 8683,
     8857, 8861, 8890, 8892, 8902, 8909, 8981, 8985, 9099, 9108, 9122, 9134, 9253,
     9278, 9330, 9353, 9354, 9360, 9395, 9412, 9424, 9426, 9534, 9558, 9599, 9605,
     9615, 9639

Water status in plant and carbon fixation pathways 8673, 8863, 8902, 9122, 9534

Water status in plant and carotenoids 8874

Water status in plant and chlorophyll 8156, 8247, 8360, 8430, 8589, 8646, 8699,
     8786, 8874, 8969, 9134, 9188, 9224, 9353, 9354

Water status in plant and chloroplasts 8345, 8861, 8865, 8874, 8953, 8969, 9122,
     9134, 9188, 9309

Water status in plant and conductance for water vapour and carbon dioxide transfer
     8131, 8133, 8134, 8189, 8190, 8221, 8224, 8239, 8252, 8306, 8360, 8383, 8387,
     8406, 8413, 8417, 8515, 8525, 8549, 8597, 8629, 8645, 8662, 8667, 8676, 8677,
     8727, 8744, 8745, 8750, 8751, 8771, 8800, 8821, 8861, 8902, 8907, 8927, 8929,
     8944, 8951, 8999, 9005, 9074, 9094, 9134, 9141, 9224, 9225, 9226, 9250, 9261,
     9273, 9278, 9297, 9306, 9321, 9329, 9360, 9404, 9422, 9426, 9494, 9498, 9499,
     9558, 9628, 9639, 9661

Water status in plant and electron transport chain 8247, 8519, 8589, 8960

Water transport in plant and growth, productivity 8330, 8376, 8377, 8399, 8433,
     8457, 8511, 8524, 8553, 8614, 8619, 8624, 8633, 8677, 8700, 8712, 8768, 8907,
     8924, 8970, 8998, 9006, 9022, 9088, 9108, 9122, 9126, 9211, 9271, 9292, 9316,
     9360, 9391, 9427, 9428, 9444, 9453, 9477, 9510, 9519, 9558, 9617

Water status in plant and leaf anatomy 8135, 8280, 8671, 8786, 8802, 8821, 8844,
     8873, 8874, 8907, 9074, 9122, 9278, 9558

Water status in plant and photorespiration 8902, 9122, 9424, 9639

Water status in plant and respiration 8216, 8228, 8298, 8517, 8598, 8633, 8831,
     8890, 9122, 9353, 9354, 9366, 9412, 9424, 9426, 9639

Water transport in plant see also Age of plant ...; Cultivars ...; Deuterium
     oxide, tritium oxide ...; Ecotypes, geographical types ...; Enzyme inhibi-
     tors ...; Growth substances, hormones, inhibitors etc. ...; Humidity of air
     ...; Irradiance ...; Mineral elements ...; Osmotically active substances
     ...; Oxygen ...; Pathogens ...; Precipitation, dew ...; Proteins, amino
     acids, nucleic acids ...; Soil moisture ...; Temperature ...; Water status
     in plant and water transport in plant

Water transport in plant, capacitances 8743, 9030, 9037, 9166, 9627

Water transport in plant, conductances 8267, 8302, 8330, 8375, 8621, 8743, 8860,
     9030, 9037, 9116, 9166, 9296, 9326, 9495, 9627, 9636

Water transport in plant, diurnal changes 8159, 8359, 8709, 8741, 8857, 8899, 9043,
     9636

Water transport in plant, methods 8359, 9358

Water transport in plant, radial transport in tree stems 8743, 9037, 9663

Water transport in plant, seasonal changes 8857

Water transport in plant, theoretical background 8302, 8386, 8401, 8621, 9030, 9288,
     9436, 9541, 9670

Water transport in plant, transport in leaf 8621, 9037, 9539

Water transport in plant, transport in other organs than above and below 8955, 9037,
     9117, 9265, 9346

Water transport in plant, transport in root 8136, 8267, 8375, 8401, 8514, 8527,
     8855, 8860, 8868, 8889, 9030, 9037, 9116, 9134, 9139, 9153, 9182, 9184, 9266,
     9271, 9272, 9296, 9436, 9495, 9531, 9541, 9636, 9663, 9670

Water transport in plant, transport in xylem, methods 8331, 8468, 9491

Water transport in plant, transport in xylem of herbaceous stem 8527, 8793, 9037,
     9573, 9576

Water transport in plant, transport in xylem of woody stem 8159, 8330, 8331, 8358,
     8359, 8527, 8709, 8741, 8743, 8802, 8824, 8855, 8857, 8899, 9037, 9100, 9139,
     9358, 9491, 9627, 9663

Water transport in plant, transport soil - root 8136, 8292, 8401, 8443, 8651, 9030,
     9037, 9134, 9153, 9266, 9436

Water transport in plant, vascular bundle structure 8328, 8375, 8523, 8802, 8855,
     9037, 9139, 9265, 9271, 9272, 9473, 9474, 9576, 9672

Water transport in soil 8236, 8237, 8267, 8268, 8269, 8472, 8494, 8651, 8692, 8777,
     9030, 9037, 9055, 9134, 9153, 9166, 9169, 9191, 9264, 9266, 9313, 9321, 9389,
     9487

Waxes see Leaf surface, waxes and trichomes; Leaf surface, waxes and trichomes,
     seasonal changes

Wettability of leaves 8369, 8370

Wilting see also Age of plant ...; Anatomical structure ...; Carbon dioxide ...;
     Cultivars ...; Drought ...; Ecotypes, geographical types ...; Enzyme inhi-
     bitors ...; Enzymes ...; Flooding ...; Genetics ...; Growth substances,
     hormones, inhibitors etc. ...; Humidity of air ...; Irradiance ...;